D0093968

Everything Matters

Ideas for Supervisors

First edition

Written by

Hollis A. Palmer Ph.D.

With illustration by
Anthony Califano

Deep Roots Publications
Saratoga Springs, New York

Everything Matters
Ideas for Supervisors

Published by:
Deep Roots Publications
Post Office Box 114
Saratoga Springs, New York

Copyright 2005
By Dr. Hollis A. Palmer

Library of Congress Number 2004195000

Printed in the United States of America

ISBN 09671713-3-4

This book is dedicated to my mentor

Gerald Murphy Ed. D.

who saw more in me than I saw in myself.

This book is also for all those who, over the years,
were my interns.

You gave me as much as I gave you. It is now your
role to pass on the ideas and beliefs behind
Everything Matters.

Special thanks to
Jennifer Brazgel
and
Sharon Mitchell
for helping me express my ideas in a logical, read-able format.

Also

My thanks to

Jim Russo

Who has made all six books look so good.

Table of Contents

People are Judged (48)- Two eyes, two ears, one mouth (56) - What have you done for me lately? (58) - Facing difficulties - Take the High Road - Stress points - The Laundry List - The Bully - How Much is enough - Sources of Support - The Ripple Effect - Finding your own bike to ride - The Feeling of Constantly Giving - Know your own demons - Working from a Draft - Time Ally - Time Enemy - Are You Having Fun? - The Worst Days - Personnel issues are never gone - Beware of the First Person Who Tries to be Your Friend - The Employee's Trap - Distraction - Popularity versus Respect - The Reason for Foresight - The Fairness Doctrine - Everyone wants to respect his boss! - A real decision is never Black or White - Praises, not Raises - Dream Stealers – Justification – Procrastination - Respect is an ever-shifting variable - It All Comes Out When you are out - Taking a Day Off at Work - Suit Day - The Hand Shake - You can never go back - Same Problems, Different Faces - Never Complain about your Salary - They always focus on the one weakness - Complaining can be a form of Boasting - There is a time to do what is easy first – Disagreements - Pass on all of the credit - The Truth is in the Details - Who's Hiding Now? - You never know who you affect

Why did you get the position? - A Different Way to Look at Candidates - Evidence of a Supervisor Who Lost - Evidence that a Supervisor was a Winner - Dress to Fill - Dress to Interview - Words to listen for - Another behavior to listen for - Never Hire of Promote a Friend - Knowing who can be promoted - Interviews Have Two Sides - When Does an Interview Start? - When does the Interview End? - A Few Years of Excellence - If in doubt, hire the brightest - The problem with an early start - No one wants the caboose gone

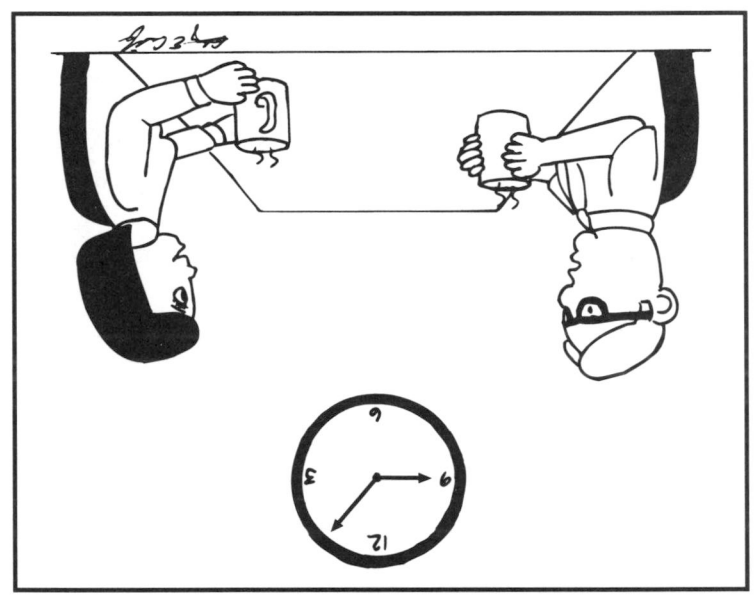

"There really are days when nothing goes right!"

Overview
Introduction and Preface

Done correctly, supervision is an art. Like all forms of art, some people are better at supervision than others. Like art, supervision can be classified, studied, described, categorized and evaluated. In supervision, as in the arts, it is through study, practice and honest critiques that a person improves.

Throughout history, the process through which people have learned the arts has evolved. In the days of apprenticeships and journeymen, people learned a trade by watching master craftsmen; at that time people learned to be supervisors from mentors (the relationship was one-on-one or small group). Today there is an ever-increasing reliance on academic study as the best, or at least the quickest, way to learn the skills of supervision. Supervision, when learned as part of an academic pursuit, allows students (in large groups) to learn from the experiences and research of others. Despite the increased reliance on formal education, most people still learn the practical aspects of supervision by observing others who are supervisors (alone).

This book is designed to help people to become better supervisors by combining the three methods of learning the art of supervision – mentoring, experience and observation. Unlike theoretical text, this book is based on real world experiences. The book is full of examples and ideas that can be related to the readers' own experiences. In fact, the reader will find himself mentally placing names from his own experiences in the examples in the text.

This book is a collection of ideas expressed as axioms. The axioms were developed by looking at the similarities among actual experiences. The decision to use axioms evolved, but was based on a belief that people remember important points best when they are reduced to short, meaningful expressions.

The book, **Everything Matters**, was designed to be used primarily by people who are or who aspire to be supervisors. The book may also be helpful for people who are experiencing difficulties with their own supervisor, since it may explain the reasons for some of his behaviors.

Research finds that we remember 10% of what we read, 20% of what we hear, 30% of what we see, 50% of what we see and hear, 70% of what is discussed with others, 80% of what we experience personally, and 90% of what we teach to others. This book provides a framework to learn by observing supervisors using specific parameters. Thus by moving from just reading the book (10% retention) to experiencing through planned observations (80%) the book increases the percentages readers will remember. In the case of readers who are already supervisors, the book provides concrete ideas to reflect on with respect to their own ongoing supervision.

Reading this book can be beneficial (10% level); however it is recommended that the book be read and discussed with colleague(s). The axioms and ideas that comprise the text were developed with the intent of opening conversations and sparking debates. The resulting dialogs will help the reader internalize the ideas presented and raise the retention to the 70% level. When reading this book in a cohort setting, if each colleague takes the lead in discussing one axiom, it provides an opportunity to examine several work-related situations. This process allows the colleagues to learn from the successes and mistakes of others (80 – 90% level).

The book contains a plethora of Break Ideas identified by coffee cups. Each of these Break Ideas could have been discussed in much greater detail; however, the ideas were intentionally limited in order to provide peers with ideas to initiate discussions.

After colleagues have read the book, they may find themselves debating which axioms apply to a given circumstance at work. This is the best possible reaction; because it means that the peers are analyzing a problem and trying to classify their perceptions. Debates are to be expected and should be encouraged when examining some of the axioms. While a debate is occurring, keep in mind that the difference in opinion is usually in how a person approaches a problem. However, if after a lengthy discussion the peers do not agree, it may be time to try to develop a new axiom that covers the situation.

How-to books, such as this, are truly effective when a reader finds himself able to relate experiences in his own life to points in the book. In the case of **Everything Matters**, proof of the validity of a section is demonstrated whenever a reader finds himself putting names to ideas or thinking of examples from his own workplace. The book has served as an instructional tool when a reader finds himself quoting the axioms as new situations arise in his organization. A reader has truly grasped and accepted the concept of *Everything Matters* when he finds himself creating his own axioms.

Aspiring supervisors should practice by taking situations that are developing at work and matching the issue to one or more of the axioms. As the situation continues to progress, the aspiring supervisor may decide that different axioms become more appropriate. These evolutions in appropriate axioms demonstrate that problems evolve and what may be a solution at one stage will be inadequate or inappropriate at a later point.

Reading the book with a peer is also recommended for people who are currently supervisors. The peer does not have to be someone from work, but should be someone at a comparable supervisory level. There is a potential problem for supervisors, who work for the same organization, when reading the book together. Since they may, at some point, be rivals for future advancements they may hold back and actually not be candid in the discussions.

In many ways, this book was designed to serve like a cookbook. Instead of recipes to make a special dish, this book has axioms to express important points in supervision. Where a recipe provides a formula, the axioms in this book are meant to provide a guideline. Where a recipe is modified to the tastes of the chef, the axioms in this book will be modified to match the skills, setting and strengths of the supervisor. A chef will not even try some recipes, a supervisor may not be comfortable using some of the axioms. In the same way that most chefs are better with some parts of a meal than others (desserts over the main dish), supervisors are almost always better with one aspect of their position than in others (hiring over evaluation). As a chef challenges himself by trying a new recipe, a supervisor should challenge himself by trying a new axiom.

While this book was being written, the fundamental concept that *"Everything Matters"* spawned a dilemma in the use of personal pronouns. Clearly, women are significant fixtures in the supervisory workforce. There was long and difficult discussion about whether this should be recognized by the use of he/she and him/her every time a pronoun was used. The problem is that in a short manuscript this combination is readable but over the course of a book becomes annoying. The editorial staff – which was all female - and author discussed alternating pronouns but it was agreed that the use of one gender in some of the sections could result in appearing negative toward that gender. It was finally decided to use the conventional grammar default "he" or "him" with a full acknowledgement that women are absolutely equal players in supervision. It is also hoped that someone will come up with a gender neutral pronoun.

Remember:

If you are thinking of examples, the book has your attention!
If you are quoting an axiom, the book has made a difference!
If you have created your own axiom(s), then you have accepted the concepts in Everything Matters!

How you got this book
Tells you a lot!

There are many ways by which a reader could have obtained a copy of this book. How a reader, who is a supervisor, acquired his copy indicates a lot about how he is perceived; after all, *Everything Matters*.

The good ways of getting the book

If the reader acquired his copy at a workshop, that is great; it means that either he or his supervisor appreciate the need for ongoing training. Remember that the book was intended to be used in conjunction with a professional workshop or seminar.

If the reader is an aspiring supervisor and someone gave him this book, congratulations are in order, for the person who gave the book wanted the reader to be successful.

If the reader bought the book, he demonstrated self-motivation, and an understanding that everyone can improve.

If the reader is a supervisor and received a copy as a gift from a friend or family member, this indicates that he has the support of the person who gave it to him. The giver sees the reader's anxieties, frustrations, and moments of gratification and wants to show the reader that these are natural outgrowths of supervision. The key factor here is that the reader knows who gave him the book.

The "not-so-good" ways of getting the book

If the reader is a supervisor and found this book on his desk, he has been given a message. Someone feels that he needs support. It may have been a member of his staff or possibly even his own supervisor but clearly someone wanted to get the reader's attention. This interpretation does not apply if the reader's supervisor provided copies to all of the people whom he supervises.

If the reader is a supervisor and found, on his desk, a copy that had an occasional page folded over and some passages marked with yellow highlights, it tells him even more. Someone feels that he needs to improve; the person who gave the book is even showing the reader areas in which development is warranted.

If the reader is a supervisor and found, on his desk, several copies of the book with folded pages and highlighted passages, he needs to remember that things could be worse. Several members of the staff feel the reader needs to improve, but those who gave the individual books have not yet formed a united group.

If the reader is a supervisor and found an abused copy of this book on his desk, with numerous pages folded over and passages marked in several colors, it tells the reader that the staff, as a group, is having problems with his supervision.

If the reader is an employee and received a copy of this book with a note attached saying, "Read, make notes, and pass it on," it tells the reader what his peers think of their supervisor.

If the reader is a supervisor and found a copy with cryptic notes in the company washroom, it means that someone forgot it there and that the option noted immediately above has not been completed. The reader should leave it where he found it; it will appear on his desk eventually.

If the reader is a middle or upper level supervisor and the employees of one of his lower level supervisors sent a marked copy, it indicates to the reader how they perceive the skills of the lower level administrator. *Help him to receive the training he deserves.*

Keep in mind that the important point is not how the reader got the book; the point is the message contained inside.

The reader should remember that if the book was a gift, even an anonymous one, it means that the giver has not given up on him.

There is only one time a supervisor has to worry about having received this book as a gift. If he missed a day of work, and when he came in the next day, he found the pages had been used to wallpaper his office, he has major problems and someone did not want him to ignore them!

Everything Matters
How it began and went forward

Analysis & Synthesis – looking for what is similar

The title of this book is based on an expression that my mentor, Gerald Murphy Ed. D., and I used over and over during the five years that we worked together. Whenever an issue was brewing – which was pretty much all of the time – in addition to considering ways to resolve the problem, we would try to analyze the source(s). Consistently, our analyses were not only of the person(s) who generated the problem – that was usually easy to determine – but also of the events or reasons that served as the catalysts for that person to create the challenge. Later we began synthesizing a series of similar incidents. When we found connecting links, we would turn to each other and say, "Everything Matters." After the first year, we started to develop and to assign unpretentious axioms to these incidents and behaviors. Later, these simple axioms served as instant reminders of similar issues and how they were resolved. During the time that we worked together, only a handful of axioms were finished; however, after Jerry's untimely death, I continued the practice in each ensuing position, in partnership with other administrators.

The process of developing axioms based on analysis of an incident followed by synthesis of that information with an analysis of similar issues proved to be a system that worked. It worked because the process calls on the participants **to look for what incidents have in common, rather than what makes them different**. The best part of the process was that, in a future incident, a match could usually be made to one of the short axioms. The axioms now served as a reminder of how a similar problem was handled.

The most was learned, in the shortest amount of time, in my second organization, where I was employed as the labor relations specialist. A labor relations specialist has intimate knowledge of the mistakes of other supervisors and their subsequent outcomes. The labor relations appointment was the only time I held a staff position (see position levels). The organization that employed me was a cooperative. It was created to employ specialists that the individual members could not justify on their own. Within this organization I could observe both line and staff positions and the people who worked in each capacity.

As the only labor relations specialist, I worked with the supervisors of twenty-six independent organizations. The organizations, which were all in the same industry, had a mixed history of relations with their respective staffs. At the time of my arrival, the employee relations in most of the organizations would be described as strained. My position was created to try to relieve some of the pressure. Many of the organizations were small, with only one or two professional supervisors; however, two organizations had over 400 employees each. At first, I was only called upon when there was a serious situation such as an employee who needed to be disciplined, a grievance or union negotiations. While in this position, there were numerous negative situations. Suddenly, the concepts and practices used in *Everything Matters* made even more sense. At first, I was never called in until the problem situation had already occurred. Since I was an outsider who was "staff," I was less vested and, therefore, could be more objective in my analysis. The specialized nature of the labor relations position provided me with a unique opportunity to observe and to analyze. During this period numerous axioms evolved by reviewing cases in which a mistake had been made by an employee, a group of employees or by the supervisor.

While in the labor position, I worked with a few supervisors who caused most of the problems. The experiences with those administrators contributed extensively to the section of the book devoted to how much can be learned by observing someone who you do not respect.

After three years of maturation and building trust, a degree of labor peace began to develop. As the trust grew, the intent of the labor services began to change from dealing exclusively with negative situations to how to attain, and then maintain, a reasonable level of accord within the various organizations. With this transition, the process of *Everything Matters* became more difficult. Now I was trying to analyze and synthesize from the practices of supervisors who were able to establish equitable labor relations.

Throughout the remainder of my career, which included working for four different organizations, the practice of trying to find common threads among problems continued and expanded. In each organization I found people who shared the desire to address problems in a broad way rather than as a collection of unique incidents. These people became my interns, as I had been to Jerry Murphy. They are not named for fear one would be left out (see section on you never know who you affect). The consistent practice of working with others dissecting problems into their component parts, and then comparing those parts to other incidents in order to synthesize a solution, perpetuated the first principle of *Everything Matters* – the principle that others are needed for each of us to reach our individual best.

As time went on, the axioms grew in number and were used in conversations with other supervisors. One of my early interns jokingly referred to the axioms as "Palmerisms", a name that stuck in all the subsequent organizations. A "Palmerism" was defined as an observation of human interaction that was not scientifically proven but showed consistent truth. Over the years, those with whom I worked contributed to the collection of ideas. This book reflects ideas discussed with many people, most of whom are currently supervisors, some of whom are not even close to their personal career peaks.

During the thirty-five years spent developing, using, and, occasionally, ignoring the axioms in this book, it was obvious that what really matters to any business or organization are its people. People are not just the backbone of any organization; they are its body and soul. The decisions made by people are the reasons that businesses and organizations succeed and fail. It is, therefore, essential to get the right people, in the right positions, with the right support. This book was designed to help address how to accomplish those three objectives. There are also some ideas about what to do if, after training, a person is still not right for a position.

Experiences in a variety of organizations outside the fields from which the axioms originally developed demonstrated that they worked in virtually the same way in service organizations, and not-for-profits, as well as in private business and governmental institutions. The axioms of *Everything Matters* proved true in all aspects of our professional lives.

The reader should be aware that even though I developed the axioms in this book it does not mean that they were always practiced. Being fully knowledgeable of what should be done, there were times when I took a different course of action. This is not said with pride, but is said to allow the reader some consolation when he slips on the wet slippery slope of supervision.

As mentioned before, this book should be utilized as one uses a cookbook; it is a collection of recipes. When a person refers to his mother's recipe, he acknowledges that it is not simply the ingredients that determine the outcome; it is also the personality of the chef. Like all good recipes, to be truly effective, the ideas in this book need the special touch of the individual who implements them.

The axioms in this book were never meant to answer all questions, nor will they work in every situation. Depending upon the circumstances and the position of the reader in his career, the axioms are meant to serve as:

- A set of guidelines for everyone
- A set of strategies for those who are just starting out
- A reminder for those who, like many of us, have fallen into a pattern
- A validation for people who adopted the practices long ago, but never wrote them down
- A way to be more successful in the role of supervisor
- A way to reexamine the performance of every supervisor.

Organizations

"No this isn't a not-for-profit. We just haven't made any money!"

The one term that covers them all.

Organizations

The decision to use the general word "organizations" rather than talking in the multiple terms of businesses, not-for-profits, service groups, clubs or government departments was based on the recognition that, although each of these entities has its own purpose, in most respects, they are very similar. The reason they are similar is their key element, which is that they all are comprised of people.

Organizations, regardless of their purpose, have:
• A personality
• Problems
• A culture
• A structure, both formal and informal
• Leaders, both formal and informal

Organizations are either expanding or contracting; rarely do they stay the same size. As they expand or contract, they are forced to face the issues created by the inertia of either direction. At the same time that an organization is dealing with the energy generated by the changes in size, it must recognize the need to evolve. Even when the need for change is recognized by an organization, there are internal forces at work to resist that change.

The seemingly complex issues relating to change and inertia at work on organizations today have been plaguing social groups since the days of the prehistoric hunts. Given this history, there is no reason to believe that the conflicts will cease as long as people continue to work together. The realistic challenge for organizations is determining how to stop wasting time trying to end stress, but rather to re-channel the energy into productivity.

Arrows Up

Arrows Down

In which direction is your organization moving?

The direction in which an organization moves is governed by the quality of the people involved. An organization moves forward through success in recruiting and promoting.

Each issue of *Newsweek* contains a section in which the magazine records changes in how the public views select individuals. If a person's image is improving, his arrow is pointed up. If a person's public rating went down, his arrow points downward. *Newsweek* also uses a horizontal arrow to designate no perceived change. To provide for a range in the amount of alleged change there are also distinctions for slightly up or slightly down.

Using the *arrows up or down* model is an easy way to determine the direction in which any organization will move. The direction and speed with which an organization is changing is determined by examining the direction of the arrows in the filling of its last five openings. Note: five is a minimum, with ten being the upper limit. To determine the direction of the arrow, consider the quality of the person who was hired when compared to the quality of the person whom he replaced. The comparison is made by answering the question: "Is the new person better than his predecessor or, by comparison, is he a step down?" The angle of the arrow is determined by the degree of improvement or decline.

Having established the direction of each new person's arrow, determining the direction of the organization as a whole is a simple matter. An organization moves forward when the preponderance of arrows are pointed up. An organization is moving down when the arrows are pointed in that direction. If the arrows are mixed, the organization is in a state of flux.

13

The other factor that needs to be considered is the importance of the position that was filled. A change in the chief executive of an organization is far more significant than a change in an entry level position. The relationship of importance to arrow direction can be adjusted by considering factors, similar to the way in which judges determine the degree of difficulty in diving, i.e. if the chief executive changes, multiply the arrow by three, if a vice president changes use two point five and for the receptionist by one.

There is one major caveat, dealing with time that needs to be included in measuring an organization's movement by the direction of the arrows. The new person should be assessed in two different time frames – short term and long term. A different grid should be used for each. The first grid is the assessment when he begins his new position. The second grid is the longer term prospects for the person. It is very possible that the newly-appointed person is not as qualified, initially, as the individual whom he replaced. However, given the lessons learned over time, that successor may become equal to, or even greatly superior, to his predecessor. To use this "arrows" model accurately, one should examine both the short and long term.

There was an organization which had a ninety percent turnover of the management staff in a five-year period. The *arrows up arrows down model* was used consistently. For the first four years, the arrows nearly always went up. These angles were consistent, using either the short-term or the long-term grid. In the fourth year, the CEO became distracted by family issues. For one year (the fourth), the staff pulled together, and the organization continued its upward trend. By the fifth year, those in the central office had become frustrated with the prolonged period of distraction and started to seek other employment. It took only one year (the fifth), for the direction of the organization to change radically. This illustrates the impact of the head of an organization on its future.

You want to be far enough ahead of the pack so that you can see when the hill is cresting before anyone else. It is better to jump off a wagon before it starts down the hill.

An Organization's Personality
Opposites attract; Likes fit in

After five years, an institution has the personality of the person who is in charge.

Organizations are like families; each has its own personality. An organization may be very professional with an "all business" attitude; another may be laid-back, while a third may be considered to be quiet and conservative. Given time, small to medium size organizations develop the personalities of their respective heads – or their most influential person. Unfortunately, too many organizations end up like families, laced with infighting, anger and frustration.

In addition to the person who is the head, there are numerous contributors to an organization's personality. Like a person, an organization's personality develops based on its appearance, the influences of other similar organizations around them, and its locations on the spectrums [see section]. Like a family, actual age as well as the maturity level of its members matter. A new organization will still be developing its personality, while an established business has an evolved character. None of these factors though come close to having the impact of the personality of the person in charge of the organization.

The time it takes for an organization's personality to be the same as its leader may vary, but usually happens within five years. If the leader of an organization is disorganized or responds from a panic (reactive) mode, the organization will respond likewise. If the head is well-prepared and methodical (pro-active), the organization he supervises will follow suit. The organization's personality is, in fact, determined from a set of expectations; which are the expectations of its chief officer (head), so the transition to conformity with his personality is inevitable.

There is, however, one clear difference between a family and an organization. Unlike a family into which a person is born, an organization recruits its new members. Therefore, an organization tries to recruit people with whom it is comfortable - like-minded people. It is difficult for a person to change his personality, and because people recruit others like themselves, it is equally difficult for a business to change its personality, unless the organization makes a concerted plan to change. By necessity, an organization goes through a transformation whenever there is a change in the person who is the head. A difference in leadership provides the best opportunity for an organization to alter direction and personality.

There are many reasons why the selection of a head is important to an organization. Recruiting services are usually used to find a new head. These services perform a series of exercises designed to determine the needs of the organization and the skills necessary in the person selected to be the replacement. It is equally important for all organizations to look at their persona and the personalities of candidates for key positions. Selection committees should try to determine whether a candidate's personality will simply "fit in" or will promote positive change in the organization.

There are at least three major considerations associated with this theory.

- What if you are the leader of an organization and you are unhappy with its personality? You may need to look at your own behaviors.
- The issue of formal and informal leadership comes into play: the correlation of personality to leader is not necessarily with the formal leader, it is with the real leader. This difference is most obvious when the formal head is more of a figurehead than a true leader.
- Does this theory suggest that, by definition, all partnerships are schizophrenic?

Leaders and Managers
The Pendulum Is Always Swinging

There are leaders, and there are managers. Both have a place, but nothing is less successful than when a leader is placed in a manager's role or a manager is placed in a role that requires leadership.

A person who has demonstrated the capacity to be a successful leader may be the worst possible manager and vice versa. Roget's Thesaurus does not even list the words *leader* and *manager* as synonyms. The two words share some common synonyms, such as "administrator," but the words denote very different characteristics. Leaders are people around whom things change. A leader may be content, but he is rarely satisfied. Managers are people who insure that the objectives of the organization are accomplished. When the operations are going as planned, a manager may be satisfied.

Leadership and management are on opposite ends of the continuum called "administrator." The center point of this continuum would be leader/manager (Note: there is no one word for a person who is successfully exhibits both sets of traits). The closer a supervisor is to either end of the continuum, the less successful that person will be when pushed into the opposite role. Conversely, the supervisor who is successful as a combination of leader and manager is rarely outstanding in either role.

The fact that management and leadership are at opposite ends of the administrative continuum causes organizations to choose one of three options. The organization can have a strong manager and be assured that the day-to-day operations will run smoothly. The drawback is that change will only occur slowly. The organization can take the opposite approach and hire a leader and watch as change happens. The drawback is that overseeing of the day-to-day operations will almost always suffer. The final alternative is to find someone who is a leader/manager. A person who tries to succeed at both roles is rarely outstanding at either. The qualities needed to perform each role are so different that anyone who tries to perform them both will often find himself frustrated and unsuccessful in either. Success in the dual role is more likely to occur when a manager is expected to perform as a leader for a short time or on a special project.

In senior level positions, businesses and governmental organizations usually alternate between the two types of supervisors. If a strong leader leaves a position, he is almost always replaced by a manager; it is as though the organization needs to take a breather. Conversely, if a skilled manager leaves a senior position, he is almost always replaced by a leader. In this case, it is as though the organization is afraid that it will fall behind the competition.

Organizations actually function best when the chief administrator and his deputy possess opposite qualities; one is a leader, and the other is a manager. Expecting people with such diverse qualities to get along may sound impossible. Actually, the closer they are to the ends of the continuum, the better they tend to work together. The relationship may be driven by the cliché that "opposites attract." What seals the bond between these two is that both recognize the need for what the other does well and that neither one wants the other's role.

Pendulums swing. A new supervisor is usually the same distance from the center of the administrator's continuum as his predecessor but in the opposite direction.

Who's Who in the Zoo?
The informal power structure

It's all in who you know.

Every organization has at least two power structures: the formal and the informal. The formal power structure is usually defined by the organizational chart. However, this chart rarely functions as designed because of the impact of the informal structure. The informal structure is uncharted and is determined by the relationships among the people inside the organization. Informal power is held by those who have the ability either to get things accomplished or to block their completion. One thing that both the formal and the informal power structures have in common is that they are usually both positive and negative.

Success is dependent upon the informal structure

Given time and consistency, the two power structures either merge or become distinct. If the supervisor is respected, takes his time to explain a decision, and knows the employees on a personal level, the two structures tend to blend, because he is taking on the traits of informal leaders. If the formal leader is perceived as authoritative, impersonal and unapproachable, an informal leader will emerge. When the relationship between the supervisor and the employees deteriorates to a point at which there is a bell (symbolic) in place to signal the employees of the boss's arrival or departure, the structures are separate and embedded.

New supervisors need to determine who holds informal power and to determine how the unofficial leader gained his influence. In most cases, the informal leader's peers understand that he has the technical knowledge and skill to be in a higher position (he knows what needs to be done and how to do it).

Every successful supervisor has established who, on the staff, has to be convinced of the viability of an idea in order for change to occur. Because the members of the staff respect the informal leader, he is usually in a critical position. The informal leader's ability to move a project places him in a position where his support is important to his supervisor. If the informal leader becomes a detractor of the supervisor, the organization will mire down. Since supervision is the process of getting the task completed, not of completing the task, it is only when the informal leader gives his support to the task that it will be completed efficiently. Therefore, it is essential for a supervisor to identify the informal leader(s) and the reasons why he is respected by his peers.

Support cannot be determined solely by the comments offered. Detractors and those who can, by their natures, see the pitfalls of an idea may sound alike in a meeting, but the former is an obstacle, while the latter is offering a necessary caution. In examining positive comments, successful supervision requires being able to discern who is truly accepting of an idea and who is agreeing simply because he rarely has a good idea of his own.

One person who had limited intelligence, minimal creativity, and no understanding of compromise, had risen to the level of informal leader in a large organization. His selection as an informal leader was a mystery until the union representative pointed out that this "leader" was single and owned a pickup truck. The "mystery leader" had helped almost everyone to move and they owed him.

Things are not always as they seem. For a variety of reasons, the informal leader may not be obvious. One person may be the informal leader and have a second person who serves as his mouthpiece. The informal leader may never demonstrate his leadership in public, working instead behind the scenes. *Supervisors need to understand that failure to acknowledge and to identify the informal power structure will inevitably create problems. Knowing how to work with the informal power structure leads to success.*

Idealists versus Pluralists
Idealists steer us; Pluralists lead us

Depending on the issues, each person has both idealistic and pluralistic tendencies. People are idealists when they hold an objective or a mission to be so important that it must not be amended before it is implemented. Pluralists, in their purest form, want what they see as possible, even if it is not perfect. To a pluralist, compromise is a part of the cost of doing business, while to an idealist, compromise is a sign of weakness.

In an organization, idealists are people who hold missions or objectives as correct and, therefore, unconditional. The idealist sees the accomplishment (implementation) of an objective as his primary purpose. An idealist would rather see no change than to accept a change that is a compromise. Idealists become so attached to the correctness of an idea that they will not waiver in trying to establish its full, unaltered implementation. Whether it is how to organize a fundraiser, or the use of software, or the dress standard of the organization, an idealist marries an idea and will not waiver. Although true idealists are rare, there are people in all organizations who hold strong idealistic tendencies. Idealists are important because they hold the organization to high standards and prevent it from accepting the easiest compromise. Idealists make an organization hold onto its core values.

In organizational terms, pluralists are people who believe that they accomplish their mission when there is a consensus which allows the organization to change. A pluralist is less concerned about the ideas themselves than about the implementation of a change. Pluralists are important because they want to get things done (implemented).

To show the effect of each, assume that an organization is considering the design of a new building. The idealist would study the various options and select a series that he feels will meet the needs of the people in the organization (**Remember that an idealist makes his ideas those of the entire staff**). The pluralist, however, wants a building that meets as many of the organization's needs as possible.

For our model assume that a building was designed with a formal lobby (to impress clients), a large break room, a conference room (to hold eight) and a meeting room capable of holding 25 people. In order to meet these requirements the staff will have to work in cubicles that are six feet by six feet (36 square feet).

Assume that the idealist has determined that a person needs 64 square feet of workspace (8 x 8) in order to be comfortable. To meet the idealist's requirement for each cubical, the organization would have to give up the formal lobby, the conference room and reduce the break room to just a preparation space.

To achieve a consensus, the pluralist would try to change the cubicles to 6' x 8', by giving up the conference room, but still retaining a reduced lobby, the break room and large meeting room. In effect the idealist would have accomplished part of what he wanted – larger individual workspaces.

The problems start when the idealist demonstrates that he will not accept compromises. If, when the pluralist offers compromises, they are repeatedly rejected by the idealist, eventually, the pluralist will stop trying to please the idealist. The pluralist will decide that, since the idealist will not compromise, then his part of a consensus is impossible. At this point, the idealist actually starts to hurt his own cause. If the idealist just will not bend, the pluralist in the organization first stops trying to get his agreement, then, later, the pluralist will actually turn against the idealist and make a recommendation that is in direct conflict with the idealist's goals – make the work stations 5 x 5. Over time, if an idealist will not compromise, he actually begins to hurt his own cause when the others stop listening or, worse yet, begin to deliberately propose ideas in direct opposition to the idealist's views.

Pluralists are most effective when they do not bend too much. Idealists are most effective when they bend a little.

Too rigid and things break, too flexible and there is not solid support.

Position Levels
Odd – Even / Line – Staff

The assistant may be the worst person to take over.

Because they are managed by people, organizations have a tendency to try to fit the infamous square peg in to a round hole.

It has been demonstrated repeatedly that the best predictor of a person's success in his next position is his success in his past positions. This reasoning usually applies to internal promotions. This choice may be good; however, examining success at position levels will increase the likelihood of making the right choice geometrically.

All positions in an organization break into two major divisions. The first division is *odd* or *even*. The second division is *line* or *staff*.

	Line	Staff
Odd	President Manager Superintendent	Technician Specialist Director
Even	Vice Assistant Aide	Associate Deputy

Odd level positions, whether they are line or staff, have a great deal control over their own destinies and, often, the destinies of others. *Odd* level positions have supervisors but the supervision is at a distance with limited monitoring. *Even* level positions, line or staff, report to someone else and the supervision is fairly close and is monitored more frequently.

Line positions fit directly into the mission and goals of the organization. *Staff* positions are specialists; they have responsibility for a specific project or activity.

Positions are usually identifiable by their titles. *Odd* level positions, that are also *line* positions, have titles like manager, superintendent, or president. *Staff* positions that are *odd* have titles like specialist or director. *Even* positions, whether line or staff, have titles with prefixes like assistant, associate, vice, or deputy.

Failure to understand the skills necessary for success in each of these quadrants results in one of the most foolish of management's mistakes – inappropriate promotions. There is an assumption that someone who is good in a position with the title "assistant" will be good in the next position up the ladder. In fact, the exact opposite is too often true. The fact that the person works well with direction or in the shadow of his supervisor will be his downfall when he is directly responsible for an office or division. The better candidate is the person who is successful as an assistant because he has the know-how, but who is unhappy because of his lack of autonomy.

In these quadrants, time of service in a position becomes a critical factor. A person who is by nature an *odd* can work for a limited time in an *even* position, assuming that he understands from the outset that it is a learning assignment. A person who is successful in an *even* position may be successful for a while in an *odd* position, based on his expertise; however, with time, his need for direction will result in job dissatisfaction.

People who go into staff positions (even if they are successful) are rarely able to return to line positions. The fact that they are specialists – project oriented – means that they like time to ponder decisions and they have a tendency to cubbyhole their knowledge. The skills that make them successful in staff positions hold them back from the quick decisions more commonly required in line positions.

If someone has had the title "assistant" for too long and he is successful in that role; watch out if he is promoted.

Fixing the Problem
Not the Blame

When trouble arises, people fall into two categories – those who try to fix the problem, and those who try to fix the blame.

There is a great story about a supervisor who, as he was packing up his office, met the person who was going to be his replacement. The supervisor said, "I left you three envelopes in the bottom desk drawer. They are numbered. Open one each time you find yourself in trouble."

A couple of months later, the replacement found himself confronted by what he thought was an impossible crisis. He remembered the envelopes and opened the first one. It said, "Blame me." The new supervisor went out and explained how the issue was the fault of his predecessor. The explanation bought him some time and the issue died down.

A few months later, a second, even more difficult problem arose. Remembering the envelopes, the supervisor decided to open the second one. It said, "Blame your boss." He again explained that it was not his fault, saying that he had done as instructed. Again the explanation bought him time and the issue died down.

As always happens, time passed, and a third, even bigger, problem arose. The supervisor returned to his desk and opened the last envelope. The note said, "Leave three envelopes for the next person."

This joke is an example of the one of the biggest pitfalls with regard to leadership. Really good leaders and managers will spend a few minutes trying to establish who or what caused the problem, but will spend as much time as necessary resolving the issue. Fault really only matters if it is repetitive or dangerous; while solving the problem allows the organization to move forward.

There are too many examples of the practice of fixing the blame. Probably the best examples are visible nightly on the evening news. Since the space program and the downfall of the Soviet Union, there has not been a national goal that was not social in nature (education, health care, etc.). Some degree of resolution to social issues may be possible, but since they involve interactions with people, the issues can never be totally solved. As a result of not having a national mission that can be completed, the political agenda is trying to address issues for which there is no one solution. Since there is no true resolution, whenever an issue arises, politicians immediately try to assign blame rather than to define a solution.

Organizations that have misguided leadership will follow the political model. When a problem occurs, these organizations focus their efforts on establishing who was at fault rather than on solving the problem. Trying to assign blame consumes valuable resources (time at the very least). If the leadership does not either change or change its mindset away from blame, the organization will diminish in quality because it devotes too much energy to finding fault.

Organizations with the right people in charge will accept that there is a problem, then set out to determine how it should be resolved. This does not mean that these organizations will not try to determine responsibility. The difference lies in the amount of resources expended on developing a solution versus assigning blame.

Long-term success is attained when you fix the problem, not the blame!

Reasons for Grievances
The Employees have reasons to speak.

Whether it is called by the collective bargaining term of "grievance" or just considered a complaint, issues that reach the formal stage between supervisors and employees happen for four reasons. There are grievances that are based on the interpretation of the wording of a provision; these are totally legitimate. There are also grievances that are valid; meaning the supervisor somehow violated the employment policies. Unfortunately, there is a third type of grievances pursued in an effort to make the supervisor look bad. The fourth type of grievances are those that are political; the employees know that their interpretation of the provision is not valid, but they feel that the issue must be put forward. Regardless of the category, grievances serve some purpose.

Employment policies or union contracts are put together by well-intended people to cover anticipated circumstances. Over time, situations arise that do not exactly fit the wording of the provision. When this happens, a supervisor usually discusses the options with other supervisors before he implements a decision. At some point, the supervisor has to make his best decision regarding application of the policy. The employees may not agree with the supervisor's interpretation, and formally file a grievance. Situations such as this are easy to identify because a logical person could argue either side. These occurrences are totally legitimate grievances. Frequently, the parties control their own destinies by coming to an agreement that covers the new situation without the use of an outside arbitrator.

The second type of grievance occurs when the administrator simply did not follow the contract or policy. In the best-run organizations, there are times when a supervisor misapplies a policy. Supervisors should remember that policies are binding obligations, and are meant to be followed. If a supervisor is responsible for a grievance, he is better off the sooner it is resolved.

The third type of grievance is usually not a single grievance but rather a wrath of actions. In a union setting, the employee group often tries to set the stage for negotiations by filing grievances on issues that will be part of the negotiations. There is a second time when filing multiple grievances is used as a weapon. This is when the employees are trying to imply that the supervisor needs help or is incompetent. When relationships between supervisors and employees are amiable, but a contract provision is not applied correctly, the issue can generally be resolved through a discussion. When the union wants to show upper management that is having a problem with a supervisor, the employees will sometimes begin to file actions. Since they are trying to make a point, the employees usually will not be willing to settle the grievance until it has been heard by the supervisor's supervisor.

The last type of grievance is the biggest nuisance for an organization. This is the case where the union knows that its interpretation of the policy is not right, but for political reasons it does not want to deal with its own member, so they put forward a grievance. There is an expression that explains the union's feelings. "We can lose this grievance but we cannot give it up." The issue is simple; the grievance is pursued so that the union can save face.

Whether called grievances or complaints, employees need a way to express concerns. Supervisors are advised to read and to be knowledgeable of the employee handbooks, employment policies or contracts. It is in employment issues that the supervisor needs to remember he is an administrator not necessarily a policy maker.

A supervisor's role is to administer the policies, not to change them.

Nine Day Wonders
The life of a crisis

Crises build to a crescendo – slowly the upsurge passes
Whether the issue is politics or the politics of a business, society is fortunate that people have a short attention span. Although a negative issue, once raised, will rarely fade into oblivion, it will work its way into the recesses of peoples' consciousness. In an organization, the life of a crisis is nine days after the last statement made by management.

"Nine-day-wonders" are crises within an organization that have the entire community talking. The most common crises of this type are associated with the dismissal or discipline of an employee or with workforce reductions.

When a crisis occurs (and one will), silence is almost always the supervisor's best choice. If management is right, take the high road and have "no comment" on the situation. Attempts to demonstrate the correctness of a decision, especially if the attempt embarrasses an employee, usually backfire. Likewise, if management is wrong, it is foolish to make matters worse; so, again, have "no comment."

There are times, however rare, when it appears to be correct to go public. When this is the choice, tell everything the first time, or else the supervisor will just be slowly feeding the rumor mill. Since it takes nine days after the last statement by management for the cycle to end, each statement prolongs the timeline. If an organization decides to go public, it needs to remember, that nipping a crisis in the bud with a one-time, clear, concise statement will save countless hours of frustration and mumbling.

One of the most practical ways to demonstrate a "nine-day-wonder" is to look at the way layoffs versus plant closings are handled. It is very likely that the news media will carry a story about "work force reductions." An announcement like this always results in a crisis in the work force. If those affected by a staff reduction are not announced at the same time as the cutback itself, the crisis will smolder, constantly rekindling itself. In the case of a plant closing, everyone knows they are out of work, and one short series of articles about the crisis is all that will surface. After the initial surge, the articles will be about the individuals who are affected and what the change will mean to them. The articles will stop being about the decision to close the plant.

In an effort to avoid feeding what would otherwise be a *"nine-day-wonder,"* there are four principles to adhere to:
- Never get in a spraying contest with a skunk
- Never discuss personnel issues
- Never discuss another person
- Silence is golden

Unless the rumor mill is fed, it will starve.

When you do good deeds;
You feel good

There is biological evidence that when a person is involved in an activity that helps others, his own health improves. When the supervisor, staff or organization has a mission of assistance, it actually makes the organization more positive and productive. *Thinking and acting positively on behalf of others helps the giver as much as the receiver.*

Critical Mass
The Amount of Energy Necessary to Create Change

When the Atomic Bomb was being developed, scientists working on the project had to actually guess the quantity of uranium (*critical mass*) that would be necessary to create a nuclear reaction.

Critical Mass in an organization is the amount of energy (attitudes and beliefs) necessary to change the vision, direction or processes of an organization.

Within an organization, a *critical mass* is necessary to refocus the energy to meet new challenges or to meet existing challenges in a different way. Like the nuclear scientists, the supervisor does not have an exact formula and must estimate the *critical mass* necessary to change the organization.

Critical mass is reached when the energy of those seeking change exceeds the inertia of those who want things to remain as they are. It is important to understand that a *critical mass* is not simply a matter of the numbers of people involved, but the energy, power, and commitment of those people.

Attaining a *critical mass* in an organization happens in two ways. *Critical mass* can be attained through careful recruitment. Recruitment as a vehicle to create a *critical mass* is risky, since what a person says in an interview and what he truly believes are not necessarily the same. The second method of attaining a *critical mass* is through convincing those already on the staff that change is necessary and appropriate. This is more difficult because the only people who really like change are babies in diapers.

The easiest way to change attitudes about an idea, thus creating a *critical mass*, is through staff development [see]. Elsewhere in this book there is a section on training and workshops; however, in the context of *critical mass* theory, the necessity is focused training, with planned representation, and structured follow up. Focused training is the process of having people trained in the desired practice. Planned representation means insuring that the informal leaders are trained and that the supervisor needs to attend the training with them to insure that the informal leader understands the concept. Structured follow up is the additional training and time spent at staff meetings focusing on the desired changes.

If forty people attend twenty different workshops on eighteen different topics, a *critical mass* will never be reached. If all of the informal leaders attend the same workshop and buy into the idea, a *critical mass* may be reached quickly.

The attainment of *critical mass* is often a function of the right "messenger." The right person devoted to an initiative can convert the staff, one person at a time. Conversely, the wrong person devoted to the same initiative may turn everyone off to the entire concept. Although it is not necessarily correct, there is usually a direct link between the idea and the messenger. The correlation between the messenger and the idea exists regardless of the merits of the proposal.

Supervisors need to understand that any desired change, cannot be attained until the idea has the support of a *critical mass* of the staff. If a supervisor moves before a *critical mass* is reached, he risks failure. If he waits too long, he will deal with frustration.

There is a second caveat to the implementation of this process. Before the change can be started, there has to be agreement among the majority of the staff, beyond just *the critical mass,* that transformation is necessary.

No one ever said change was easy, only that it was necessary and requires a critical mass to accomplish.

Temporary Promotions
are Not Foreplay

Never make someone an interim; if he wants to be considered for the position, it only teases him. The lesson in this section relates only to temporarily assigning a person to a supervisory position; the same is not true when a person is actually promoted.

There are circumstances that, when they backfire, make one realize that the situation should never have arisen. This lesson is repeated over, and over, when organizations place a staff member, who may be interested in a promotion, in a temporary supervisory position. The logic behind the temporary promotion is, "Let's give him a try." The idea is that the trial period will somehow serve as an audition. On the surface, a trial period seems sensible but it creates a totally false situation.

The person in the interim position is on trial. Suddenly, he is not a true supervisor, since, if he is not successful he may go back to his original position, nor is he a staff member because he may continue as a supervisor after his trial period has ended. The interim has found himself in an undefined place. The interim is torn. In ways he sees the opportunity as a tryout, a chance to prove himself to the organization – like a tryout for a team. The situation, however, is almost the total opposite of trying out for a team. In a team tryout, the options are, to prove yourself or you are gone. In a temporary assignment, the options are to prove yourself or to go back to your old position. Being gone and going back to the old position are not same thing. In an interim position, the person feels loyalty to his peers with whom he may be back working alongside. The old loyalty makes the interim uncomfortable with confronting the people who he is expected to supervise; after all, he may end up working with them again. Since he will not confront his peers, he is not truly able to show how he would perform as the permanent supervisor. What is the result of the interim period? The organization does not accomplish the original goal of giving the person a try at the real job.

Since the trial period rarely works in the way it was intended, the organization is then faced with one of the hardest tasks, telling someone, who thinks that he did an acceptable job, that he is not getting the permanent position.

If an organization is sure that it is going to promote a staff member to a supervisory position, the organization should demonstrate its belief in the person and just make the promotion without an interim period.

If an organization is not going to promote a staff member to a supervisory position, then the organization should not give that impression – do not even interview the inside person.

If an organization is not sure whether it is going to promote an inside staff member to a supervisory position, then the organization should have an open search and select the best candidate.

No one ever said that the majority was right
But We Want It!

One of the problems with living in a democracy is that people have the perception that the will of the majority should rule. Governments may need to respond to the desires of the majority, but that is not how effective organizations should operate. Everyone in an organization may feel that he deserves a benefit (i.e. to have off the day before Thanksgiving) yet, unless the entire industry adopts the same benefit, the organization that grants the day will become less competitive. For a supervisor, in cases in which a decision provides a direct benefit, there is "right" and there is "popular." If a supervisor stays with "right" his long-term chances for survival are far greater than if he goes for "popular."

People with the same Title
The founding of associations

Associations can be viewed both cynically and optimistically. A cynic would hold that associations have a limited purpose, adding that they begin by meeting after work, but later evolve into meetings during the day (taking time from work).

An optimist would say that associations provide a time for people to discuss common problems. The fact is that both views are correct.

In organizations where there are few supervisors with similar titles/responsibilities these individuals find that associations provide an essential support mechanism. Members of associations provide advice, counsel, or validation to a fellow member. One of the biggest obstacles for small organizations to overcome is the lack of peers for their supervisors; associations help fill that void.

If three people, from two or more organizations, have similar titles, they instantly form a formal association.

An Underlying Problem with Downsizing:
People working in positions that they were not hired to fill.

There is a long-term outcome to downsizing and hiring freezes that is often overlooked. When there is an economic downturn, organizations tend to keep as many of the current staff as possible. During downsizing and/or hiring freezes, the work has to be assigned to those who remain. With people not working at tasks for which they were hired, the organization, which is already struggling, is now faced with the added costs for retraining and from lost productivity. There is a far greater long-term problem from hiring freezes. Since the organization cannot recruit, it is forced to fill supervisory positions from within. While the freeze is on, people are promoted who would not otherwise be considered. When the freeze is over, those who were promoted will retain their positions, and the company will be forced to fight the *Two's Hire Three's* Syndrome.

The POS-sition
As a supervisor, one of the most discouraging situations involves dealing with a peer who wants only the trappings of the position but who does not want the responsibilities. These are the supervisors who want to come and leave at will, to go to lunch, to park in a reserved spot, to wear a suit and to attend

meetings, but are not committed to fulfilling their supervisory responsibilities. These people are posing in the position or they only want the POS-sition. These people are easy to identify at a meeting; they are the ones who take all of the credit and pass all of the blame.

Luck is a Combination of Opportunity, Preparation, Attitude and Willingness to take Risk

There are people who appear to be luckier than others – they appear to be consistently in the right place at the right time. Some of this may, in fact, be just luck (opportunity), but more frequently it is the result of other factors. These people have prepared themselves for a chance to improve (preparation). They have completed the requirements for the next opportunity before the door was ever open. These people are almost universally optimistic (attitude). They may not be overly verbal about their optimism, but their positive actions and attitudes are recognized by those who make decisions. When any opportunity finally presents itself, there are risks involved. Opportunities bear a cost. The cost may be job security, money or time, but "lucky" people understand that without risk there is little to gain.

There is a formula for being lucky. Be _optimistic_ while _preparing_ for an _opportunity_ which will present itself, then take a reasonable _risk_.

Investments versus Expenses

When any resource (time, money, materials) is consumed, it is an expense. An expense is the consumption of a resource for which no return was expected. If, in the long-term, an expense is expected to return its value, it was not an expense; it was an investment.

During the real estate boom of 1999 – 2004, people spent money on their homes for such things as interest, utilities and taxes. If a person bought before the rush and sold at the peak, he would have realized that every cent he had spent, including interest, utilities and taxes, was returned. What was considered, at the time, to be an expense was, in fact, an investment.

From an organizational perspective, staff development and training are often considered to be expenses. Viewed as an expense, training is among the first budget lines to be cut in difficult times. The truth is that staff development demonstrates the organization's commitment to the staff members who participate. Since salary rarely satisfies an employee, individual recognition by encouraging a person to grow professionally is one of the most important investments for any organization.

Negatives are Specific
Positives are General
When listening to comments, made by the staff, it does not take long to realize that there are far more negatives than positives made about a supervisor. The reason is that negatives are very specific and, therefore, more exist. "Did you see the tie he had on?" While positives are usually more general "He always dresses so well." This is why, if one out of three comments that are heard about a supervisor are positive, he is either winning or at the very least in balance.

Confusing the Message with the Messenger
It is not just what is said it is who says it.
One of the biggest dilemmas facing organizations is to avoid associating the quality of a cause with the person who champions the idea. An idea brought forward by an unpopular employee does not necessarily make it a bad idea, any more than a suggestion brought forward by an informal leader or popular employee is automatically good. The problem for an organization is trying to implement a good idea that is put promoted by an unpopular person or in preventing an unwise idea that is sponsored by a popular person.

Supervisors should always follow policies
They were the boss's ideas of fair.
There is an important reason for an entry or middle level supervisor to follow policies and the contract. The contract or

policies were approved by the heads of the organization. The head agreed that the policies would be the rules of the organization. When a supervisor misinterprets or fails to follow a provision, he is, in effect, not doing what his boss agreed was fair. When this happens, the supervisor is replacing his superior's judgment with his own – rarely a wise move.

It's not who you know

It's who knows you and how well you are doing
Many people make the mistake of believing that being acquainted with important people will somehow help them in their careers. Most assuredly, there are cases in which knowing "the right people" did help a person's career. Much more frequently, people are assisted in their careers, not by the people they knew, but by the people who were familiar with them and their performance. The difference lies in who cares most about the relationship. The person, who thinks that he "knows" someone important, is the one placing the value on the relationship. If person "A" claims to know person "B", then "A" values the relationship and should not necessarily count on "B" for help. If instead, "B" knows about "A" then "B" values the relationship and may support person "A's" career. The situation is about as easy to accept as the last two sentences were to read and understand.

The reality is that *who knows you* is often more a detriment than a benefit.

Supervision

"So, is it true, the boss has found fourteen copies of the book on his desk?"

Supervision more than a game; less than a victory

Supervision
The art of helping others do their work.

As soon as there are two people working together, there is a degree of supervision. It would be logical to assume that the amount of supervision is directly related to the number of workers; however, anyone who has ever driven with a backseat driver in the car can testify that sometimes one-on-one supervision is the closest and the most annoying. Conversely, when a supervisor is responsible for numerous employees, he actually spends less time supervising each. As the number of people for whom a supervisor is responsible grows, employees who perform at average or above, find that they are virtually ignored.

Beyond direct oversight, there are numerous other aspects of supervision including; development of realistic expectations, staff assessment and time management. Supervisors are less frustrated when they find and communicate the balance between reasonable expectations and contributions. When each person contributes at a fair level, based on his ability, a supervisor's mission is accomplished. Fair and candid evaluations of each staff member are essential elements of the supervisor's professional success. Supervision, especially as it relates to evaluations, is sometimes a delicate function. Tactfully telling people how they can improve, as well as in what ways they have helped the organization is an essential proficiency. It is, after all, as unfair to expect too much of a person, as it is to accept too little.

One of the fundamentals of successful supervision is using time wisely. Whether it is spent in meetings, in analysis of problems or in managing frustrating staff members, time is a valuable resource. Time is a precious asset that needs to be expended prudently.

Evaluations
When only the truth will do

Too often a supervisor demonstrates his own limitations by failing to provide members of his staff with what they deserve – a forthright assessment of their performances. Failure to address a staff member's failings may, in part stem from a belief in one of several clichés: "If you have lemons make lemonade," "You can't teach an old dog new tricks," or "You can't make a silk purse out of a sow's ear." No matter which one is used, it becomes an excuse for failing to perform one of the most important responsibilities of a supervisor – providing a person with a chance to improve. If, in his mind, a supervisor complains about an employee's performance, he owes that person an opportunity to correct the deficit.

Each organization needs to establish a process for staff evaluations. Numerous resources exist to help with staff evaluations, including books and workshops. Regardless of the specifics that an organization develops for evaluations, each supervisor should be trained in what to look for and how he should express his sentiments. For difficult evaluations, the supervisor needs someone to serve as a sounding board insuring that what he wants to say is acceptable before he reviews the evaluation with the employee.

Evaluations, written or oral, tend to fall into three categories; an incident summary, a summary of the performance on a specific activity or a performance review. The purpose of each is different, but each must provide the person with an unbiased assessment and the opportunity to improve.

Of the three types of reviews, the incident summary is the most likely to be negative. These summaries say that on __ date __ happened. A supervisor writes an incident report either to congratulate the employee on handling a situation well (rare) or to counsel the employee on how to handle a similar situation in the future because the situation in question was not handled well. The key elements of incident summaries are honesty and clear explanations of what needs to be corrected and how to make the necessary changes (follow up). If training is suggested in the evaluations the summary should indicate who will be responsible for finding the training (always the employee - the supervisor has enough to do). Failure of an employee to follow up on the suggestions in an incident report, including training, can be grounds for his losing his position.

When a major project is completed, supervisors commonly complete an assessment of each staff member's contribution. The very nature of the report (completion of a project) causes the supervisor to have a positive perspective. Supervisors are reminded that what they write can be used against them later. If someone did not make a contribution, do not give him credit. Supervisors have two good choices; either to send general memos congratulating everyone who was involved, or make very specific assessments of each person's contribution. Taking a shortcut and trying to reuse the same wording on each memo will usually backfire because the people who received it will make comparisons and realize that the memo really does not give recognition for individual contributions.

Annual performance reviews are the perfect forum to address employees' issues and contributions. The reviews should be set up to measure performance against the job description – actually going through each line of the job description is a good place to start. Using the job description prevents supervisors from being unfair by criticizing someone for something that was not his responsibility. In all annual evaluations, be honest and announce that all employees will receive at least one suggestion for improvement. As a supervisor, remember that if you do not state a weakness, it is unfair to expect improvement.

When it comes to staff evaluation, only the truth will do, and everyone deserves to know what he can do to improve.

Chip Theory
Every decision is a gamble

Decisions may be a gamble, but the fall-out from some bets (actions) is more predictable than from others.

Supervision is not a game, but understanding how to "keep score" is essential to being a successful supervisor. Two of the biggest problems in supervision are that the rules are modified by each person and that the value of the chips is always changing. Those two variables make it imperative that every supervisor understand the process of supervision.

The premise of "Chip Theory" - Supervisors hold a separate stack of chips to be used with each employee. At the same time, each employee has a stack of chips to use with his supervisor. Chips are moved based on actions that are taken, usually initiated by a request or a decision. Each request or decision has the potential to move one or more chips. Chips move in only one direction; they are never traded. A chip may be earned by the employee and given by the supervisor or vice versa.

Rule 1: Each action has its own chip value, but the same action may have a different chip value based on those who are involved. An example of "Chip theory" in action involves the granting of time off. Given time off, one employee might owe his supervisor several chips because his reason for wanting the time was unwarranted or urgent. A second employee, given the same time off, may only assign it limited chip value, since he really didn't need the time or believed that he was entitled to the time in accordance with the personnel policies. A third employee might actually want a chip, since he only asked for the time in order to placate a significant other and was secretly hoping that the supervisor would say, "no."

Rule 2: The same action can cause chips to move from more than one employee. Granting time off to allow a staff member to see a doctor may result in many chips from that employee to the supervisor. At the same time, chips from the person's peers could be lost if they believed it was an act based on favoritism.

However, if the person who took the time is seriously ill, and the supervisor is performing a humane act by not charging the time to accruals, and is not asking for anything in return, that supervisor may actually pick up chips from the person's peers.

Rule 3: Every time a supervisor asks something of one employee at least one chip changes hands. Every time an employee requests something of the supervisor, at least one chip changes hands. Actions that do not follow past practice have a value of their own.

Rule 4: The number of chips that change hands based on an action is determined by the value ascribed to the action by the person who made the request and, at the same time, the value is determined by the person who granted the action. Since both parties establish the chip value of the action, the two parties usually see the value at different levels. The further the request is from the "norm," the greater the difference. Something that is reasonably common within the organization will have an established range in terms of chip value. An unusual request has no established chip value; so, the difference in value assigned by the parties will often be significant. In most situations, the person receiving the chip(s), whether it is the supervisor or the employee, ascribes a higher value to the action than the person who had to give up the chip(s).

Rule 5: Disagreement over the chip value of a decision will always result in problems. If the person who asked for the action or event and the person who granted it do not believe that it has the same value, there will be problems. If the difference in perceived value is minor, neither party may even be aware of the issue. If the value of the difference is major, the parties are not seeing the request from the same perspective. There is no judge to rule on the chip value of an event, so both parties can only hope that a future incident/decision will even out the values of the chips.

Rule 6: If the entire staff is told to accomplish a goal, then there is little or no movement of chips. Since everyone was treated equally, there is no exchange of chips. Chips are only exchanged when an individual or small group makes a request or has a request made of them.

Rule 7: If, subsequent to the initial meeting in which the entire staff was told to accomplish a task, an individual employee requires follow-up, there is chip movement – Application of Rule 3. Any subsequent meetings with an individual, even if they regard a consistent application, fall under Rule 3 because the meeting was with an individual.

Winning at Chip Theory

Both parties, employees and supervisors, either win jointly or lose jointly at Chip Theory. Over the course of time, the number of chips held by each party changes based on decisions or requests that have been made. As long as the piles of chips between each employee and his supervisor are nearly even, the relationship works smoothly. Relatively even stacks of chips indicate that the relationship is balanced, respect is mutual and fairness is the standard; in effect everyone is winning! One side can never win at Chip Theory; it can only dominate.

Losing at Chip Theory

In Chip Theory a loss occurs when a person or group has significantly more chips than the other. There are two ways in which uneven distribution of chips creates a losing situation. One occurs when the chips are uneven with an individual employee; the other occurs when the chips are uneven with a group of employees. Uneven piles of chips are a clear indication that one party is becoming "beholding" to the other. Too often in the workplace "beholding" leads to a dominant/subservient relationship. The reason for this outcome is that no one, neither a supervisor nor employee, likes to be indebted to another. The loss relationship is true whether the person with more chips is the supervisor or the employee.

Losing with an individual

If an employee has considerably fewer chips than his supervisor, he has asked for appreciably more than he has given. When an individual employee is rapidly depleting his chips, there are four possible outcomes:

- The best outcome is that the employee tries to replenish his chips by some action or set of actions; usually that is accomplished through improved performance.
- The employee quietly finds another position and gets a new pile of chips with his new supervisor.
- The employee will do something foolish and needs to be dismissed, thereby cashing in his last chips.
- The employee will try to rally support from his peers, borrowing from their chips.

Conversely, if an individual staff member has a larger number of chips than his supervisor, he has quietly become dominant. Unless played carefully, the employee usually loses, despite the fact that he has the additional chips. Supervisors resent a sense of obligation. In order to maintain their authority, they will attempt to balance the chips through some set of actions. The only actions that will balance the piles of chips are those which require that the employee yield (lose) some of his chips. Since the employee can only exchange a chip by a making a request, there is no way for the employer, who is behind, to balance the piles of chips, through his own initiative. Note: If the employee boasts, in some form, about his collection of chips, it is virtually always detrimental to the employee.

Losing with a group

A second way to lose at chip theory is for the supervisor to hold appreciably more chips than a large group of employees, or a large group of employees to hold more chips than the supervisor. Dominance in either direction creates unrest in the workplace.

A group of employees, who are down in chips, even if it is for different reasons, will find support in each other. This sense of mutual obligation will unify the staff, resulting in opposition to the supervisor's actions. In a union environment, this imbalance will result in grievances, prolonged negotiations, distrust and hostility. In a non-union environment, the unrest will be less visible and more indirect, resulting in rumors about the supervisor, the rise of an informal leader, and a lack of loyalty.

Conversely, on occasion, the staff, as a group, finds itself in a situation in which their individual piles of chips are higher than those of their supervisor. When this imbalance occurs, the supervisor becomes powerless. In this situation, the best outcome for the employees is if the powerless supervisor accepts he has become ineffective. The other situation exists when the supervisor tries to collect a lot of chips through a few bold actions. Since this person is unwisely behind, it is doubtful that he can construct a positive way to balance the piles. The only legitimate reason for a supervisor to be this far behind would be a medical or a family emergency.

Winning and Losing at Chip Theory

It is possible to both win and lose at chip theory at the same time. This is done at the individual or small group level. The supervisor can have even piles with virtually all of his staff, except with one who is low in chips. When this situation occurs, it is time to balance or to sever the relationship.

Caution 1: There are times when large numbers of chips are ventured at the same time. If a professional supervisor loses his or her temper in front of a group, that supervisor might lose many chips. If an employee demands a raise and threatens to leave if it is not given, he is venturing all of his chips on the one action. The same betting of all chips on one action, is true when someone is dismissed. This process of betting all the chips on one action is extremely common when one of the parties has very few chips left because the venture is really of little value.

Ways to improve at Chip Theory

There are ways for a supervisor to improve at Chip Theory, although it is "technically" not a game.

- To avoid disagreement over the value of a decision, discuss its importance with the employee. "How important is __?" or "Why do you need __?"
- In every organization there are people who are more willing than their peers to help resolve issues. Supervisors too often look to them for relief. Rather than making the chip piles uneven by consistently asking one

employee to complete an activity, spread the requests throughout the staff. This is the fairness doctrine.

• On a regular basis (weekly), supervisors need to evaluate the placement of chips between themselves and each employee. The supervisor should determine what chips were changed and where. The sooner any disparity is discovered, the easier it is to balance the piles of chips.

***

Key Concepts

"Chip Theory" is not a game, but it is an excellent measuring instrument.

The only thing certain in "Chip Theory" is that the chips are always moving.

Organizations function best when the supervisor is ahead, but only by a little.

Author's note: Chip Theory could be a book by itself (who knows, maybe someday it will be.)

People are Judged,

In part, by the company they keep.

Whether it is done formally or in quiet conversation or in total silence, people are still judged by criteria that are not considered to be politically correct. One bias that consistently exists is the assessment of people by those with whom they choose to associate. The concept is that people choose their friends. This form of assessment may not be fair or right, but it is still done.

Supervisors should ask themselves, how would their staff judge them based on their associates?

Doing What is Necessary
Instead of What is Important

The only way to find a path out of the forest is to step back from looking directly at the trees.

The vast majority of supervisors find themselves busy from the moment they walk into the office each morning until they leave at the end of the day. The people they see and the tasks that they undertake during the intervening hours are all required and have a bearing on the performance of the organization. These supervisors leave each day having done as much of what was *necessary* as possible, but usually having missed most of the *important*.

The optimistic perspective is that the very fact that a supervisor spends all of his time on the *necessary* prevents him from doing what is really *important*. The pessimistic perspective on this phenomenon is that a supervisor does not know how to do what is *important*, so he focuses on what is *necessary*. Experience has shown that if each supervisor would spend 10% of his work time on determining what is important and addressing it, he would actually gain 25% more time.

Most supervisors need to develop a process whereby they are more efficient. Addressing each problem as it arises does not allow time to think. Through reflection, problems are analyzed and similarities found. Once the common threads are established, the supervisor can develop a resolution that will handle more than one problem, thereby gaining time because he is not repeating a solution.

At first this process sounds like either a lot of verbiage or a fantasy, but the concept is sound. Supervisors need to make time each day to analyze and to determine common threads among the problems that they have handled, in the last day. Common threads are usually found in the areas of systems, communications, duplication of effort or personnel. There may not be a common thread connecting every problem to another, but there will be common threads among some of the problems.

After establishing the common threads, the problem should be defined in one sentence, i.e. lack of understanding of new software. To be sure that the definition works, the supervisor should wait a day to see if the definition still holds. The next step is to develop three or more potential solutions. The reason for multiple solutions is to avoid overlooking a better resolution because action was taken on the first option considered. Once the solution to these multiple, connected problems is implemented, it decreases the number of times that the supervisor is called upon to do the *necessary*.

The process of dealing with the *important* is easy to understand and to accept as a business solution; however, it is difficult to implement for three reasons: learning how to determine the common threads, finding the time and dealing with the personnel involved.

Finding common threads requires analysis and synthesis. This is best learned by utilizing a team consisting of people who are comfortable discussing issues and options.

The most common reason for failing to utilize a process of addressing the *important* actually makes sense. If the supervisor does not have any extra time now, where will the ten percent for analysis come from? There are no simple answers except to *make* the time. Sometimes the process has to begin at home or during the daily commute. Dealing with the *important* works, and the time spent on it becomes an investment, not an expense.

Too often supervisors avoid the *important* because the issue is personnel. On too many occasions, the common thread is an individual; however, one should always begin the analysis by assuming the problem is not a person. If the individual who is the problem works for the supervisor, he needs to be dealt with – he is *important*. If the person who is the common thread is a peer, or worse yet, the supervisor's own supervisor, the problem and solution become trickier. In the latter case, simply having clearly defined the problem often provides some relief.

Time spent on the important is an investment. Time spent on the necessary is an expense. Invest in yourself.

People Only Question Things
That They Think They Understand

Keeping them in the dark vs. being prepared

In general, people do not like to appear to be imprudent. Few things make a person look more foolish than asking an inappropriate question in a group setting. Therefore, people will only ask a question about a topic that they think they understand. This self-imposed limit on the topics that a person will question is an important concept, since it can be used to a supervisor's advantage or, if ignored, can create problems that could have been avoided. Supervisors should be aware that this self-imposed limit applies equally to those whom he supervises and to those who supervise him and it is present in every organization.

When dealing with a supervisor, a fellow employee, or even a friend, people do not inquire about areas in which their lack of knowledge makes them feel vulnerable. This explains why, when people ask politely, "How are you?" they really want to hear, "Okay" for the answer.

People are universally afraid that it they ask a question, the person of whom they are asking a question might respond with a question. This fear is acquired in school whenever teachers ask, "What part of (whatever the topic) didn't you understand?" Assume for a minute that you were required to attend a lecture by two Nobel Prize winning chemists on the effect of some compound on developing cells. After sitting for forty-five minutes, being exposed to terms you had never heard before, the professors ask if there are any questions. In all probability you would sit in your seat with your hands tucked under your thighs, too baffled to even consider raising your hand.

Most people cannot understand a position they have not held. Since most staff members have not been supervisors, they do not understand the responsibilities of their supervisor. It follows that employees can only question the performance of their supervisor on the most basic of his duties.

There are some general employment traits that everyone

accepts, even if they do not agree on their relative importance. All supervisors will, to some extent, be questioned on these general qualities. Examples of general employment traits are: punctuality, length and frequency of breaks, courtesy, promptness of work, fairness, respect, leaving early, appropriateness of dress, etc.. Because these values are almost universal, successful supervisors understand that those whom they supervise are judging them on these criteria. Failure at any of these traits is easy for a supervisor's staff to measure.

A supervisor has two options for addressing his staff's lack of understanding of his role. He can inform the staff of what he is doing, and later keep them abreast of details and progress. (Note: if he tries this approach, he will falter at some point and then be accused of failing to keep the staff fully updated.) The other option is to leave the employees in the dark. By never trying to explain what he is doing, the supervisor will not be accused of failing to keep the staff fully informed (he never tried); rather, he will be judged solely on the universal traits of all employees.

The same road, other direction

Ultimately, supervisors are supervised by someone else; even the CEO has a Board to which he reports. As an employee, the supervisor needs to be sure that he is prepared to deal effectively with his own supervisor(s). This is accomplished through a supervisor's awareness of his supervisor's areas of strength. Background will play heavily into the areas in which the supervisor's supervisor has expertise. Example; if the supervisor has a strong financial background, he will question costs and margins. Remember that supervisors are like everyone else; they want to feel that they are contributing. As an employee, accept and encourage the supervisor's suggestions. At the same time, as an employee, the supervisor should anticipate as many questions as possible.

To be properly recognized, each employee needs his supervisor to understand how well he is doing.

Those Who Challenge
And those to Challenge

Below is a grid that groups employees according to their intelligence and their energy levels. People sometimes ascribe percentages of staff members who fit into each of the four quadrants; however, a supervisor only needs to understand what are reasonable expectations for the people who fall into each of the four quadrants. (A "bad day" increases the percent of people in quadrants with the higher Roman numerals.)

	Bright	**Dull**
Ambitious	I	IV
Lazy	II	III

The staff members who are included in Quadrant I are those employees who are both bright and ambitious. Unless somehow corrupted, these people are the guaranteed winners or *one's* (see *One's hire One's*). By their very natures, people who fall in this quadrant will define problems, develop and consider solutions, and make necessary adjustments. A supervisor's role is to keep these people well-supplied, challenged and satisfied. The supervisor needs to be sure that these employees have both the necessary materials and the training required to improve. To these people, supervisors should be sure to pass on all the credit.

Staff members in Quadrant II are the supervisor's greatest challenge. These people are bright but they are lazy. These people often procrastinate and perform at, or just above, the minimum standard; these people are the *two's*. People in this quadrant can do the work, but they do not want to do the work. The fastest way for a supervisor to improve productivity is to increase the production of people who fall in this quadrant. The people in Quadrant II need motivation (sometimes that is a threat).

The staff in Quadrant III are dull and lack ambition. With this group change is a problem. Any process that is new or that is different from what is customary causes them frustration and confusion. These people are the *three's*. The best that a supervisor can do with people in this group is to allocate to them carefully-prepared assignments that are as systematic as possible. These people work best when they can learn that one step consistently leads to the next step. Because they lack ambition, these same people need constant supervision. The largest number of people who fall under the *caboose effect* can be found in this group.

The staff that falls into Quadrant IV are those people who are dull but are still ambitious. These are the people who cause a supervisor to have nightmares. Because they are dull they make mistakes. Because they are ambitious they make lots of mistakes. In short, these are the people who just keep screwing things up. Supervising people from this quadrant is like holding a tiger by the tail. If the supervisor lets go, the supervisor will be bitten by one of their mistakes. If the supervisor holds on, he will eventually become exhausted and frustrated. The only answer is to cut them loose. In this quadrant the *caboose effect* is not true because the people cannot be retrained.

Some people don't need to be fixed, some people need a charge, some people need to be monitored and some people need to be released.

Walk and Talk
The Way to Shorten Meetings

Throughout organizations, regardless of whether they are governmental, private, not-for-profit or proprietorships, "Too much time is spent in meetings," is a comment echoed in the halls.

The question becomes: how can a supervisor reduce the amount of time spent in meetings?

There is a story of a CEO who read the book <u>The One Minute Manager</u> by Kenneth Blanchard and Spencer Johnson and took it to heart. Allegedly, he put a one-minute timer on his desk. Whenever a person put his head in the door and asked, "Have you got a minute?" he would say, "Yes," and start the timer. This practice would certainly reduce the length of meetings. The problem that presents itself is that most great ideas take more than one minute to explain.

A better way for a supervisor to reduce the length of a meeting and still have the option to extend the time, if appropriate, is to adopt the practice of *Walk and Talk*. The opening line becomes, "I was just going to grab a cup of coffee; why don't you walk along with me, and we will talk about <u>(the issue or problem)</u> _." This same line can be used if a person shows up at the supervisor's door unannounced. Instantly, the supervisor has put a time frame on the meeting – the time it takes to walk to the coffee machine and back. Additionally, the supervisor has provided an agenda for the meeting (the issue that he stated). By putting a time frame on the meeting, the niceties are eliminated and the time is focused.

By using the *Walk and Talk* method, the supervisor controls the time (of the meeting) based on the value of the discussion. If the discussion can be brief or if the employee is unprepared, the meeting ends after the coffee is poured, and the two return to their separate offices. If the idea just needs a little more time to explain or refine, the supervisor can stand near the coffee pot and talk longer. If the idea appears to be developed, but requires further discussion, the supervisor can invite the person into his office to finish the discussion – yielding instant control over time and agenda. *This does mean, however, that the supervisor cannot keep a coffeepot in his office.*

A spin-off of the idea of *Walk and Talk* is for the supervisor to become the morning greeter. Some supervisors who work in organizations in which people arrive over a limited time span (i.e. schools, stores, banks, insurance companies) develop a practice of being near the door as the staff arrives. In addition to improving any punctuality problems, the practice makes the supervisor more visible – everyone gets to see him at least once each day.

The practice of being the morning greeter greatly reduces the amount of time spent in meetings. As employees learn to expect to see the supervisor every morning, they will learn that it is an excellent opportunity to ask that 'quick question.' Meetings held during this time are very short. The employee understands that the supervisor wants to get back to greeting everyone, so he gets directly to the point. If the employee needs privacy, or if it will take longer than a few minutes, he can use the morning greeting to schedule a meeting with the supervisor in his office.

A practice of taking thirty minutes to greet everyone may seem as though it adds work time a supervisor's day; in fact, it usually reduces the time spent in meetings, and is cost efficient.

The simplest way to shorten meetings is to move them out of a room.

Two eyes, two ears, one mouth
We were born with the right ratio.

There is an expression: "We are born with two ears, two eyes and one mouth; and that is the ratio we were supposed to use them." The message is clear. It is better to be silent and let everyone assume you are thinking, than to speak and let them know you are not. When examining supervisors (and even other staff members) one of the first signs that they have problems with relating to others is that they talk too much.

A silent person may not be popular but he is rarely disliked.

Looking Brilliant
Letting people try their own ideas

When staff members are allowed to try their own ideas, suddenly the supervisor looks brilliant just for saying, "Yes."

There are supervisors who give the appearance of not being very busy; their employees are active (but not stressed) and the department is efficient. What is the key that these supervisors hold? They have learned how to unlock the door to success by encouraging the staff to be creative, by asking seven questions for each new suggestion, and then by answering with, "Let's try it."

People who are capable of doing their jobs will design ways to perform each task better. When capable employees are given an assignment, they tend to do two things. First, they will start the task in the way that it was done in the past or as they were directed. As they continue, they will look for ways to be more efficient. Depending on the task, its complexity and their comfort level with the supervisor, they may even try a new method before they bring it to their supervisor's attention. Believing in themselves and in the receptiveness of their supervisor, they will then suggest that the new method be adopted. At this point, the qualities of the supervisor drive the outcome and the environment of the organization.

A supervisor who is only a manager has the mantra, "If it isn't broke, don't fix it." While a supervisor who is only a leader says, "Let's give anything a try." To all other supervisors, the idea will need some justification [See *Leader/Manager*].

In all likelihood, if the employee who is actually performing the task suggests a way to do it better, the chances are extremely likely that the suggestion will work. For the supervisor, the process at this point should be to follow a formula of asking questions according to the time spectrum [see *What Time is it*]:

- How did we do it in the past?
- How are we doing it now?
- What are the troublesome aspects of the current

process?
- What is the process that is being proposed?
- How (or why) will the new process be more efficient?
- What obstacles exist to implementing the new process?
- What problems will the new process create?

The result: "Okay – Go with it."

At this point, a bizarre phenomenon takes place. Because the supervisor allowed the employee to try his own idea, the supervisor is considered to be talented.

There are numerous stories of supervisors who built their entire careers on allowing their staffs to try their own new ideas. These "brilliant" supervisors almost always followed a supervisor who had been resistant to change.

There is merit to adopting a career strategy in which, for the first three months in any new supervisory role, the only changes made are those that are suggested by the staff. After all, they are the ones who have to do the job and who want to do it as efficiently as possible.

What have you done for me lately?
People Have Short Memories

Among the long list of things that causes frustration for supervisors is the constant need of some staff to be taken care of. Even worse, there appears to be a direct correlation between the need to receive favors and the person's memory. He will remember what he saw as a favor for someone else for years but forget, within days, one granted to him.

People who think in terms of "What have you done for me lately" will never be satisfied – don't even try.

Think Politically
Act Logically

The question is not: "Does the organization have political behaviors?" (They all do) The question is: "Are political behaviors controlled, tolerated, supported, or endorsed?"

All organizations have varying degrees of political intrigue. Some organizations thrive on internal politics, while others endure their existence. The supervisor will need to determine if the politics of a choice can be managed and at what cost to the organization.

To a large extent, the tolerance of organizational politics is governed by the decision making process of the person who is the chief-executive. The more he appears to make decisions based on favoritism, rather than make decisions that are logical, the more others in the organization will behave in the same fashion.

There appears to be a direct correlation between the amount of politics within an organization and the amount of outside influence on the organization. Charities are a prime example of outside influence. Since the survival of charities is dependent upon donations and fundraisers, the constant need for money allows some individuals to derive additional influence based on the contributions which they are responsible for obtaining.

Supervisors need to establish early on how they are going to act toward the political environment in their organization. For people, whose services can be terminated (everyone except the owner), to act politically (the granting the favors) will almost always eventually backfire. The old expression, "Let no good deed go unpunished," applies to decisions made for political reasons. Many, perhaps most employees, like to talk about the influences within their organization; while at the same time, they do not want their own supervisors to act politically unless the decision will be in their favor. A bigger dilemma for a supervisor is that the political machine of every organization is too powerful for even the best supervisor to control. In short, if a supervisor tries to act politically, he will almost always lose. Acting either logically or using the fairness doctrine, a supervisor may also lose; however, the loss will be slower and his actions will be justifiable.

The best example of acting politically versus acting logically is seen in the actions of any President following a disaster. Within days of a tragedy, the President will be at the scene. While he is at the scene, the first response personnel, police, fire etc. are using valuable time protecting him, rather than doing what is logical, helping those affected by the event.

When it comes to politics even the best eventually lose. Looking at the Presidencies of Richard Nixon, Gerald Ford, Jimmy Carter, George H. Bush, and Bill Clinton, it is apparent that even men who were capable of climbing to the top of the political ladder eventually lose credibility. Therefore, if the best at politics eventually lose in the political arena, supervisors should assume that, if they act politically, eventually they will lose.

Remember when dealing with politics:

Every supervisor in every organization should assume that everyone in the organization is related to someone else in that organization. Therefore, an action taken with respect to one person will be known by everyone else. Political loyalty (having done a favor) is a shallow pool that only flows one way.

Every supervisor should assume that all of his actions are being taped – a fact that may be true.

The best a supervisor can do is to explain a political decision, while he can justify a logical one.

Will these suggestions help a supervisor to win? No. They will simply stay his execution.

Scratch a Supervisor – Find an Employee
Who Does He Identify With?

How deeply do you have to scratch a supervisor to find the employee lurking underneath?

The implication of the question is simple; does the supervisor consider himself to be an employee who is responsible for his own work, or as a supervisor who is responsible for the performance of others? In other words, does the supervisor see himself as management, or does he identify more closely with the staff whom he supervises? Keep in mind that in certain situations nearly everyone sees himself as an employee (i.e. comparing effort with reward).

There is a transition period when a person first moves from staff to supervisor. Officially, the promotion occurs quickly; however, the adjustment in the person's own mindset invariably takes longer. He needs to switch from assessing his success by the amount of work he has completed, to measuring his performance based on his ability to get work accomplished by others while, at the same time generating a harmonious work environment. It is one thing to know one has the authority to tell others to complete a task; it is another to believe that the others will do the task because they were told to. There is a fundamental difference between a person intellectually knowing he is a supervisor and emotionally accepting that he is a supervisor (that is when he and the job are one).

If a person is fortunate enough to change organizations or departments at the same time that he moves into supervision, his new staff will probably accept him as a supervisor before he truly perceives himself in that capacity. After all, the staff never knew him as a colleague.

The most difficult transition is when a person's first supervisory position is within his own organization or, worse yet, in his own department. In this circumstance, he is suddenly supervising people who were formerly his peers. These were the people with whom he joked, shared secrets, complained about his predecessor and, if he is human, inevitably did something against the rules – and some of the people who were his peers, and are now his employees, remember all the details. In this case, there is a double transition. The new supervisor needs to change his mindset, and so do those he is now supervising. It often takes a strong stand on an issue to solidify the change in perception.

In general, staffs like the idea that internal promotions are possible. Promotions contribute to recognition that there can be a career within the organization. What the staff often misses is that when a promotion occurs, they are required to change their view of a "friend."

One additional problem is that there appears to be a direct correlation between how long the person worked in a non-supervisory capacity and the length of time it takes for him to see himself as a supervisor. The longer one measured his job in actual hours and quantity and quality of work performed, the longer it takes to adjust to the role of supervisor. In fact some supervisors never complete the transition.

There is always the issue of a supervisor's associates. When an internal promotion takes place, old relationships are not erased. Friends usually still want to see themselves as a friend, and the line between friendship and respect is often delicate. There is also the probability that a person who was promoted inside had issues with one of his peers. The new supervisor will need guidance in how to deal with that person now that he is his boss. The advantage of promoting from within is evident; the person who is promoted is already familiar with the organization. The difficulty in promoting inside the organization is that it is much easier to find the employee hiding just below the surface of the supervisory title.

There are some questions that help to identify how a supervisor sees himself:

• Does he regularly tolerate inappropriate staff behavior

even when it should not be excused?

- In a disagreement over the ability to complete a project, does he unreasonably take the side of the staff?
- Does he keep supervisory hours, or does he come and go at the same times as those he supervises?
- Does he talk about compensatory time?
- How does he relate to other supervisors?
- Does he have the support of his own supervisor(s)?

A promotion may take minutes; mentally accepting it will take much longer.

Facing difficulties

A concern before it becomes an issue;
an issue before it becomes a problem;
a problem before it becomes a crisis.

There are supervisors who are so skilled that they rarely have to deal with a crisis. Those same supervisors appear to have it easy, with only a few problems developing in their areas. The reason is simple; these people have mastered the ability to identify potential difficulties when they are only issues. This supervisory competence is predicated on the acceptance of a hierarchy of difficulties. The levels of difficulties in ascending order begin with *issues*, which if left unresolved, will become *problems* which, if not resolved will turn into a crisis.

The earlier potential problems are recognized, the easier they are to resolve. By dealing with something at the issue stage, the supervisor has more time, more options, and less "damage" has been created.

Take the High Road
The view is better.

In studying military engagements, officers are taught that the advantage goes to the side that has the higher terrain. In organizations, taking the higher level is universally beneficial for supervisors. In almost everyone (supervisors included) there are events that brings out the desire for revenge. For those supervisors who cannot resist the desire for payback, remember that vengeance is a dish best served cold. Supervisors who have had long careers know that containing their anger and not responding in the way that people expect, is almost always the best option. When examining people who are admired (note the word "admired" not "respected"), it is clear that they do not become immersed in petty issues. People who are admired have learned that a person who takes the high road is ultimately the winner.

When supervisors take the high road, they learn that the view is better, the air is clearer and that people are looking up to them, not down at them.

Stress points
The limits may differ but everyone has them.

Several years ago, there were a series of articles related to how events (both positive and negative) in a person's life led to stress. That stress, if not in control, resulted in negative effects on the person's body. The list of stressors included many major events such as buying a house, having a child, entering or leaving a marriage and loss of a spouse. Each change was assigned a point value. The theory of specific point values disappeared because people argued over the accuracy of the values assigned, instead of focusing on the probable relationship between stress and health.

As a supervisor, try to keep abreast of the significant changes in each staff member's life. Although stress may not be avoidable, awareness does allow for compensations or, at least, considerations that can be made in the expectations of the person's performance at work.

The Laundry List
Something that is okay for home but not for work.

There are behaviors which very quickly make a supervisor look incompetent to his entire staff. The first behavior is failing to deal with issues in a timely manner. The second behavior is to introduce an "old" issue (anything more than a month old) into a current discussion.

Building a list of issues that are all brought out at the same time (laundry list) implies to the recipient that the supervisor is creating a case against him. People can deal with one issue (note singular) when it is current. Although no one wants to be told he was wrong, he can handle it better when the facts are fresh. Dealing in a current time frame also allows the person to correct his behavior, if that is what is necessary.

A list of issues is only appropriate as one of the last steps in the long stairwell to ending the employment relationship.

When a supervisor waits to bring up an issue he looks weak, unsure and vengeful. Unless the person is going through a personal crisis, it is hard to think of a situation where waiting is appropriate.

If there is a problem, deal with it; no one looks good carrying old baggage.

The Bully
He is everywhere.
A bully is anyone who tries to intimidate others by exerting dominance. He does this by using what he sees as his strengths. Everyone acknowledges the existence of the physical bully, but bullies come in at least four other forms. Bullies use intellect, economic status, authority or the victim's emotional weakness to achieve their objectives. No matter which mask he wears, under the laws protecting employees in the workplace, a bully is a hazard for an organization. Supervisors are required to protect their employees from the influence of the bully and they need to be sure that they do not become one either.

How Much is Enough
Executives are paid based on the value of their possible mistakes
Despite the laws of economics, a person's compensation is rarely based on supply and demand (ten minutes in the break room will show that most people believe that they could head the organization). Similarly, income is rarely a function of education or experience. Where entry level positions pay more rea-

sonably – schools, engineering, and government, the salaries of supervisors rarely double those with matching experience who did not move into supervision.

The heads of television or radio networks, movie companies, and publishing houses, like their counterparts who are heads of businesses, make multi-million dollar decisions. If the decisions are wrong, it could result in the business ceasing to exist. Their compensation packages are high because of the potential value of their mistakes.

The same is true of professional athletes and college coaches. The rewards for the members of a team that win a national championship, both in direct compensation and in product endorsements, are astronomical. Therefore, professional athletes and coaches are paid not to get hurt or to lose – their possible mistakes. The simple fact is that the presidents of the colleges whose football teams are ranked in the Top 25 are not being paid as much as the head coach of the team at that college.

The same phenomenon is true of people who are self-employed. Entrepreneurs taking limited risks, such as those owning a small business (cosmetologist, owners of delis or pizza shops), may make a living, but they rarely make a fortune. The person, who risks millions buying property or starting a "hot" club, has risked a lot and has the potential for very high rewards. "You have to invest money to make money."

Since most people work for someone else, there are two outcomes from this theory. The first is, people who are making a living working for someone else will rarely make extensive money. Secondly, and more importantly to supervisors, since people will never be content with their compensation, the supervisor needs to help them to take pride in their performance. That way they are earning job satisfaction if not monetary fulfillment.

Sources of Support

Supervisors all need support and not just in sports
Within the structure of an organization, supervisors gain a feeling of support from three different directions: from above, from below and from the side. The term "support" for a super-

visor, means that he feels that his decisions will be backed by the given party. In order to be effective, it is essential that a supervisor believe that he can take a chance because others in the organization endorsed (support) his ideas. The feeling of assurance, from those above the supervisor on the organizational chart, allows him to take reasonable risks and to move his portion of the organization forward. A second direction a supervisor gains support (collaboration) is from his peers, (the side). This assistance creates the dialog that allows his ideas to be fully explored before their implementation. Supervisors also need a feeling of being braced by those they supervise (from below). Support from his staff allows a supervisor to believe in himself and in his own fairness.

The problem for organizations develops when a supervisor thinks that he is losing support in any of the three directions. The natural reaction to a loss from one direction is to try to reinforce oneself by gaining support from another direction. A supervisor loses the support of his peers, if his peers believe he is undeserving of the position; that he "borrows" another supervisor's idea; or he cannot be trusted. If a supervisor loses the ability to collaborate with his peers, he is forced to "go it alone," thus missing important dialog. If a supervisor feels that he is not being braced by his staff, he will turn to his supervisors for support, usually leading to actions even more detrimental to the employees. When a supervisor feels that he is losing support from his supervisors, he turns to his staff for support, almost always setting precedents that are not in the best interests of the organization.

When a supervisor loses support everyone loses!

The Ripple Effect

The further one is from the boss, the less likely the waves that he creates will have any impact.
There are two ways to measure distances in an organization. There is the physical distance; people whose offices are located in Miami are less concerned about a new CEO if his

office is in New York City, than if his office is two floors above their own. The other distance is measured in the number of steps on the organizational chart between the CEO and the employee. It takes a long time for an employee at the base level to feel the impact of a new CEO, while the vice presidents will probably begin to feel the impact even before the new person starts. With either measure distance provides shelter.

It is interesting to note that most upper level supervisors use two philosophies to determine who on their staff is assigned which offices. People whose offices are closest to the CEO are either totally trusted and the CEO wants their advice quickly or the people are not trusted and the CEO wants to keep an eye on them.

As a supervisor, watch where you place staff, and watch where you are placed.

Finding your own bike to ride
You are what you do in your free time.

Supervisors, regardless of level, experience or age need to find an activity outside of work that provides them with some degree of solace. Many supervisors believe this is the role of their families. A family may be a release, but too often even members of the family constantly expect a little piece of the supervisor. When this happens, the family becomes too much like work. The next thing the supervisor realizes is that his family, even though they try to support him, has become emotionally draining.

Through an outside activity, a supervisor can be away from demands. The activity can be almost anything, but activities work best when there is a degree of physical requirement and/or there is a clear completion (woodworking or gardening). Supervision is never done, so concluding something, even as a hobby, feels good.

Successful heads of organizations universally have some activity in which they participate. The same behavior is suggested for the middle level supervisors and managers. It is these escapes; these diversions, that free the supervisors' minds to examine other alternatives.

Too many readers, expecting time to do as a person wishes may seem greedy; however, failure to have a release usually results in some form of illness.

Know your own demons
They are in all of us.

It is equally as important to know a person's strengths as it is his weaknesses. Supervisors should also know and salute a third set of characteristics that each person brings to the work environment. This set goes beyond just strengths; it includes his special qualities (calmness, supportiveness, thoughtfulness). It is the fourth set that each supervisor needs to be aware of and control in himself. These are his demons. When they rear their ugly heads, demons can cause a lot of misplaced anger, misguided blame and unnecessary intimidation.

Demons (unlike weaknesses) are traits which can truly hurt ourselves or others. Demons, when under control, are only weaknesses. When they are not restrained they are decidedly different. Demons appear in many forms, from the obvious, such as the use of drugs, to the less obvious, and often suppressed, seeking of vengeance. Demons may be small and hidden, or they may be obvious even to the casual observer. The most common demon for a supervisor is ego – the idea that his ideas are significantly better than the ideas of others. A false ego in a supervisor is always followed by a feeling that he can never give the impression that he was wrong.

The beauty of demons is that it is not necessary for a person to share them with others. However, in order to attain professional growth, it is essential to control personal demons. To do that, each person has to admit his demons to himself.

Working from a Draft
There is only one way to get to consensus in a reasonable time.

When a supervisor is required to put something in writing that requires a consensus from his staff, he should work from a draft. The best process is to have two meetings. The first is to solicit input. Then, draft a memo incorporating the ideas from the brainstorming meeting. Hold a second meeting at which the staff works from the draft; the process will be much faster.

Whenever a group tries to write anything from scratch, it is a tedious, unproductive process.

Time
Time is the supervisor's ally

Fortunately, there are few true emergencies that occur in any organization. Emergencies are those instances in which the supervisor does not have time for forethought, and therefore, he is required to give an immediate response. In a true emergency the wrong decision could have a devastating outcome (i.e. the building is on fire). Since now most organizations have developed emergency plans, even most incidents that would have been considered a crisis no longer require immediate decisions.

Given time, more options are explored, and a better decision is made.

Time is an ally; as a supervisor, use your allies. Given time, the right idea almost always presents itself.

Time
A supervisor's enemy

When a supervisor misuses time by taking too long to make a decision, he demonstrates weakness, lack of conviction or makes himself appear to be deficient in authority. Supervisors misuse time when they procrastinate, waiver, avoid making a decision or avoid being at work – long lunches, golf outings.

A supervisor who misuses time for too long will find that it becomes his enemy and he will have little of it left.

Are You Having Fun?
It should not be all work.

Complaining about one's job is a tradition passed on from parent to child. There is, however, a very real difference between complaining about a position and really not liking it. If the supervisor truly does not like his position, everyone who works with him sees it. Since the supervisor sets the standards, if he is unhappy, chances are great that the staff as a whole will be discontented.

Life is too short not to enjoy what one is doing with one third of every day.

The Worst Days
The problem with an open calendar

When starting out supervisors often suffer from a false impression that, if there are no meetings scheduled, that it will be a good day. Not having anything scheduled just means that the supervisor is not given the opportunity to anticipate the issues. Not anticipating issues is not the same as having no issues.

Experience shows that an open calendar is like sitting in front of a target.

Personnel issues are never gone
They are just placed in remission

As personnel issues arise, they require attention. The reality is that each person who comes in contact with an organization, whether as an employee, customer, investor or service provider, brings his issues into the workplace. The fact that people bring problems with them is neither bad nor good; it is simply true. Eventually all supervisors accept that the biggest challenge in their role involves dealing with personnel issues. The best that a supervisor should expect is for outside issues to be placed in remission while people are at work.

Experienced supervisors know that they will never solve all issues but they are happy whenever one is resolved.

Beware of the First Person Who Tries to be Your Friend;
He probably needs one.

When a supervisor, or anyone else, is new to a position, he is almost always greeted by those already on the staff. Inevitably, one of those people will immediately try to develop a friendship. Far too frequently, that person is trying to befriend the new supervisor because he lacks friends of his own. Unfortunately, there is probably a good reason why he does not have friends among the staff.

The Employee's Trap
Do not get caught playing his game of distraction

Employees and supervisors work under two different sets of rules. The weaker, more mistrustful, or more resentful the employee is to the supervisor, the more he will want the supervisor to be distracted from his responsibility of direct supervision. Employees who feel vulnerable will try to distract their supervisors by complaining about things that are related to the workplace, but not to the mission of the organization, i.e. the tissue in the washroom or the condition of the parking lot. Whenever a supervisor is distracted by these non-mission related issues, he is not dealing with his "important" task – direct supervision. The best advice is to assign the correction to the person who is in charge of that problem. If the employee continues to make similar complaints on a regular basis, conduct a comprehensive performance review. The chances are extremely high that he does not want the supervisor to find out that his performance is substandard or, at the very least, not at the level that he is capable of performing.

The supervisor's role is to focus on getting the task completed, while the weak employee's role is to keep the supervisor off-track.

Distraction - the best weapon of the weak!
Using (Misusing) Policies

Distraction can also be used in another way. Those who are jealous or weak seem to take delight in repeatedly pointing out applications of the organization's policies that, they contend, demonstrate favoritism.

The practice begins with the disgruntled employee providing examples of applications of policies that he sees as unfair.

The examples he gives are provided in two ways and for two different reasons. If the employee is trying to show that the supervisor is perceived as weak in the organization, he will give examples of a policy application that appears more favorable to another department in the same organization – thus challenging his supervisor to even the score. The second way in which policy application is used is to try to undermine the supervisor within the employee's own department. This is accomplished by citing examples of applications of policies by the supervisor that the discontented employee perceived to be unfair i.e. "He is allowed to take longer for lunch than anyone else." In either form, the employee is trying to show the supervisor as weak and to distract him from his real mission.

Weak employees have found distraction to be the weapon of choice!

Popularity versus Respect
Which is more Important?

If a supervisor wants to be liked first and respected second, he will be neither. If a supervisor wants to be respected first and liked second, he may be both.

The Reason for Foresight
Knowing when to get off

A supervisor always needs to be in a position that is far enough ahead of the pack that he is the first to see that the hill (organization) is cresting – he does not want to be caught on a runaway wagon on the downhill slide.

The Fairness Doctrine
You have to prove right, but you only have to be fair.

Only idealists think that they can make decisions, affecting other people that are right or wrong. The best a supervisor can hope for is to make decisions that are fair to as many people as possible. Even if a supervisor tries to be right in his decisions, he will be thought of as wrong by some on his staff. As time goes by, and more decisions are made, the number of people who think that he is wrong will constantly grow. At some point, those who think the supervisor is wrong will become a *Critical Mass*. If the supervisor says, up front, that his job is to be fair, not right, he can hope to be thought of as fair or just.

In supervision, fair wins over right every time.

Using the fairness doctrine allows the supervisor to sleep at night; doing favors causes acid reflux.

Everyone wants to respect his boss!
It's as simple as that.

If an employee does not respect his boss, after a short time it reflects badly on the employee. The unspoken question is "Why are you (the employee) staying, if your boss is unworthy of his position?" In short, how can someone respect the employee if he will work for someone whom he does not respect? Care should be taken not to confuse lack of respect with complaining about the supervisor, since it is a given that people always complain about their supervisors. The difference between a complaint and loss of respect is the way in which the words are said.

Employees who do not respect their boss for too long lose respect for themselves.

A real decision is never Black or White.
It is always gray

In organizations a decision can rarely be reduced to being either black or white. In fact, when a decision can be thought of as simply as black or white, it was not really a decision, since there was no decision (everyone automatically would take the white). In reality, every decision has positive and negative ramifications; in effect, every decision is a shade of gray. The question becomes; *how close to white does a supervisor expect or how dark a gray is he willing to risk.*

Praises, not Raises
People gain more satisfaction from
praise than a bonus.

Amazing as it may seem, people get a greater sense of job satisfaction from praise than from a raise. Salary (raises included) is more often a source of dissatisfaction, since the employee regards the raise as insufficient.

Praise costs nothing, demonstrates appreciation and can be repeated.

Justification:
A Waste of Time

Whenever a supervisor justifies something not being done he is rationalizing his lack of motivation. Whenever a supervisor justifies something being done he is searching for an excuse. In either case, justification results in wasted time. The supervisor should remember that the reason a person asked for justification is because he is seeking someone to blame.

Procrastination:
A Thief of Time

A supervisor's choices are to think about a task once and then complete it, or think about it each time it is postponed. Delaying a required chore takes much longer than to simply complete the task. Ultimately, the chore has to be done, and each time it is reconsidered takes more time. In an effort to keep their consciences clear, some supervisors will complete other tasks. No matter what the excuse procrastination takes time.

It is always better to do the task than think about it – unless it is a criminal act.

Respect is an ever-shifting variable
Climbing a hill or falling off a cliff

A barometer rises and falls based on climatic conditions; respect for a supervisor rises and falls based on the manner in which situations are addressed. The daunting aspect of this phenomenon is that, with the exception of a true crisis, respect rises more slowly than it declines. While it may take years to build respect that same respect can be eroded in just days. There may come a point at which the erosion has gone on for so long that it has created ruts too deep to ever rebuild. In this case, the work environment needs the severance of someone.

The good news is that, with few exceptions, all staffs want to believe in their supervisors.

It All Comes Out When you are out.
You never know what the maid will find.

Supervisors can manage problems and keep issues, even major ones, restrained while they are at work. It is when they are absent that everything that they were managing to suppress explodes. Since the supervisor is not present to handle the situation, the supervisor's supervisor has to step in. Suddenly, others know the problem the supervisor was controlling. Even worse, those who were in attendance handled the problem in the way they perceived suitable, which may not be the way the supervisor would have wanted to resolve the issue.

There are problems with the axiom: *it all comes out when you are out.* To avoid being out, supervisors tend to hold back on their own training. Having to deal with problems keeps supervisors at work, which at first glance might seem like a good thing; however, everyone, especially supervisors, needs training, and training takes time. Unresolved issues also cause stress. Suppressing issues, which a supervisor can often do while he is at work, is a form of procrastination (a leading cause of stress.) It is accepted that stress causes health problems. Health problems cause the supervisor to miss work. When the supervisor misses work, the problem will come out and his supervisor will suddenly know about the procrastination. Since at some point the issues will have to be dealt with, why not resolve them now and avoid the hassle?

Dream Stealers
They are everywhere

Dream Stealers are those overly-vocal pessimistic people whom supervisors are forced to deal with every day. No matter what the problem, they are sure it cannot be resolved. To a *Dream Stealer*, if a person received a promotion, he did not deserve it; if a person receives an award, someone else has already performed the new task in a better way. *Dream Stealers* belittle every opportunity or success. Since people are judged by the company they keep, supervisors should think about the time they spend with those who try to rob pleasure from other people's successes.

Taking a Day Off at Work
Informal resolutions

What does the supervisor do when he takes a day off at work? He spends the entire day going to the employees in their working space. To take a day off at work, the supervisor sets his schedules so that no paperwork will be completed, and all phone calls will be returned at the end of the day. Then he spends the entire day visiting and engaging employees. Note: he visits, but does not invade the employee's space. During the day, he takes breaks in the break room, stops at the printer and listens to the staff. As the supervisor talks to his staff, in their spaces, he will hear issues that may not come up at meetings, or that the employees may not have otherwise brought to the supervisor's attention. It works and supervisors who try the process soon do it on a regular basis.

Taking a day off at work allows the supervisor to address questions before they become issues.

Suit Day
A day suited to complete the necessary.

A spin to *taking a day off at work* is suit day. There comes a point when the supervisor needs a day just to do paperwork and those other "administrative" functions. One section of the book talks about how people dress based on their plans for the day. Creating a "suit day" provides a suitable option (pun intended) for getting a lot of tasks done in one day.

"Suit day" started on a personal level. Whenever I had a very difficult meeting or was going to have to deal negatively with an employee, I always wore a dark, pinstriped suit. Going in on a day like that, I knew the day was going to be difficult, so it usually appeared to the staff that I was distracted and not in a good mood. One morning I wore "the suit" just because it had not been worn in some time. Early that morning, the administrative assistant who worked outside my doors said to someone, "Are you sure you want to see him today? He has *the suit* on." The person responded that it could wait until the next day. Apparently, "the suit" sent a very "professional" ("leave him alone") message. That day everyone gave me space, and more paperwork was accomplished than ever imagined. After that, when I changed positions, one of the first things I always told my assistant was that when I wore the pinstripe suit it was not a good day to meet with me. Now, without hurting anyone's feelings, I had a way to get caught up on the mundane tasks without being rude!

They always focus on the one weakness
No one likes criticism
There is an interesting observation that has been demonstrated to be universally true. If ninety-nine percent of the performance review is positive and one percent is negative, the employee will want to discuss in detail the one percent.

The Hand Shake
Getting a grasp on someone else
Handshakes send many messages. From who offers his hand first, to the strength of the grasp, a handshake sends many messages. Some say that the angle of the extended hand demonstrates dominance. The further over a person turns his hand, the more dominant he is trying to be. If the person turns his hand so that the back is virtually on top, he is very dominant. If a person turns his hand so that the palm is up, he is more submissive. If a supervisor believes these theories, he can learn a lot about people by extending his hand exactly vertically and seeing what they do with it!

You can never go back;
So don't even try.
Too often, people accept a position as supervisor in a new organization, with the intention of returning to their original

organization as a supervisor. This aspiration is especially true if the position in the new organization is the person's first supervisory experience. Such expectations are possible, but returning brings into play the issues of history, friendships, commitment and loyalty.

A person may go back, but the reality of the sacrifice will be greater than the fantasy.

Same Problems, Different Faces
You can run, but they will follow
The most difficult part of supervision is dealing with personnel issues because they are never-ending. The longer a supervisor is in a position the more the issues of one person or a small group of his staff begin to plague him. Inevitably, the supervisor believes that, if he were able to change positions, these problems would go away. The reality is that the problems will stay. The best that he should hope for is that the faces will change.

Never complain about your Salary
Or personal expenses
In every organization, the staff believes that they know approximately how much the supervisors earn. Whether they are right or wrong is not the issue; the issue is that they go to their humble homes and dream about what bills they could pay off if only they had the supervisor's income for just one month. These employees have not learned that, to a large degree, everyone spends based on whatever he earns. The last thing that the employees want to hear is their supervisor saying that he cannot live on his income. This same restriction applies to expenses. Employees may accept that the supervisor has new furniture, but they do not want to hear about any problems associated with the purchase.

If you make more than the person you are talking to don't complain about money!

Complaining can be a form of Boasting
It all depends on how it is said
Is the statement, "He expected me to get all ___ completed before I left, and I was still out of here on time" a complaint,

or is it boasting? The answer is possibly some of each, but it was definitely a boast on the part of the employee. If a supervisor heard the quote, he might assume that the person was complaining about his assignment; however, in all probability, the employee was bragging about how much he had accomplished.

There is a time to do what is easy first
A way to get your message out
Experience has provided one suggestion for the order performance reviews are conducted. Begin with several of the people who will do well. Even though these first reviews are positive, they need to be honest. The chances are, that those who are receiving good reviews will want to know how to improve (this is why they are doing well to begin with). After a couple of positive reviews are completed, take on the tough reviews. Using this sequence provides two positive outcomes. Since the supervisor is the writer, his frame of mind will be better after having completed the positive assessments. Second, employees whose performance reviews were weak will start complaining to their peers about the quality of the supervisor's assessments. Those who received a good review either have to agree with the complainer and believe their own reviews are not accurate (who wants to take a positive and make it negative?) or accept that the supervisor is correct and the complainer deserved the negative comments.

Disagreements
Logical, Emotional or Physical
Disagreements come in three forms: logical, emotional, and physical. Physical disagreements in the workplace are, in their own way, the easiest to resolve because they can never be tolerated and warrant a swift resolution. Logical disagreements are more difficult because they need to be weighed, and a decision made based on the quality of the reasoning, in conjunction with the needs of the organization. The emotional dis-

agreements, because they stem from beliefs and feelings, last the longest and are rarely ever resolved.

Pass on all of the credit
Accept all of the blame.
A supervisor is responsible for getting the job done. If it is not completed on time, the fault (blame) is his. If it is completed successfully, others did it (credit). In either event, the staff knows the contributions that everyone made. People on the staff may brag about their input in a project, but in their hearts they know how much credit they deserve. The more a supervisor passes credit on to his employees, the more respect he will receive from them. The more a supervisor accepts blame when things go wrong, the more people will respect him. Resentment develops whenever a supervisor accepts credit, especially for something for which he was only partially responsible. Even more resentment arises when a supervisor passes blame on to those whom he supervises.

Humility is the best ally of a supervisor.

The Truth is in the Detail
The way to tell which story is true.
There will be times when members of the staff have conflicting memories of a situation. Virtually always, there is a way to determine who is correct. It will be the person who provides the most detail (assuming that the details provided remain consistent).

Who's Hiding Now?
The one who is guilty is not in sight.
When a situation arises between two staff members or between a staff member and his supervisor, the two usually try to avoid each other. The interesting outcome for the supervisor, is that the person who believes he was in the wrong – note the person does not have to actually acknowledge that he was wrong just believe it – cannot be found. The guilty person is in his space doing his work. When it is time to make up, the guilty

party will plan something for the entire staff or bring in something for the break room.

Crossing Five Bridges
Distance creates expertise

There is an expression that a person has to cross five bridges before anyone thinks of him as an expert. This need for distance is based on the fact that within his own community the "expert" is known as a "person." Meaning that near "home" people know the expert's human qualities, especially his weaknesses. Knowledge of a person's weaknesses, which could be in an entirely different aspect of the person's life than his expertise, is the reason why virtually no one is considered an expert in his own organization. The need for distance in order to be considered an expert is not a new phenomenon; there is even a Biblical reference to inability to be a prophet in one's own land.

Acceptance of the five bridge theory is essential when a person believes that there is someone within the organization who is as knowledgeable about a topic as the so called "expert" who was brought in to lead a seminar.

Personnel

"If one's hire one's and two's hire three's, how did you get a job?"

Hired, Fired and the Steps in Between!

Personnel
It's all about people

Personnel is a catch all term that includes hiring, firing and all the steps in between. Key among those steps is evaluation. The "steps" include the set of actions designed to help each staff member perform his task more efficiently. Unfortunately, personnel is the component of supervision that is usually squeezed in when there is no direct supervision required. Later, when an issue surfaces with an employee, everyone wonders why it was not dealt with sooner (the personnel function).

By selecting the best person for a position (one of personnel's major responsibilities) the work of a supervisor gets easier. There is a team process that increases the probability of hiring a quality person. The problem, however, is that organizations often settle for an under qualified candidate. When an organization accepts the best **available** (rather than just the best) it increases the supervisor's workload. The gap between the best person and the best person available is the difference between a reasonable workload and overloading a supervisor.

For a supervisor, another component of personnel, involves his own career expectations. A supervisor or aspiring supervisor needs to accept a position that meets his style, personality and experience thus providing the opportunity for success.

Supervisors who realize that their job is all about people move personnel from a task done when time allows to a primary responsibility.

Hiring what is needed
There is a better way

The person who gets the call offering him a new position can be reasonably certain that he is the person the organization liked. Regrettably, he can be less sure that he is the person the organization needs or wants.

Recruiting parallels dating. In the end, all people want to be with someone they like, even if that person is not the best for the position (be it a job or a partner). During the screening process, organizations consistently move from trying to find someone with the skills it needs, to seeking a new person who is liked. The dating analogy provides a model for screening. If people were asked to list ten traits needed in a partner, they could make, and probably even prioritize, a list of qualities that they felt were important in a long-term relationship. Placed in a pool of eligible partners, people soon drop the list of needed traits and start looking for someone they like (physically, emotionally or both). The new dating services understand this phenomenon and have changed the process slightly, allowing them to legitimately claim many successful relationships. Dating services have adopted an additional step; they only match their members with people who meet the "need" criteria. After the initial screening, members are selecting people whom they like out of a smaller, but more appropriate pool. It's similar to trying to pick a winner in a 10 horse race when it can be determined that only two of the horses have a chance of winning.

In recruiting, priorities should be sequenced: **required, needed, wanted, liked, desired and bonus** traits, the same progression that dating services use for screening potential partners. Additionally, there needs to a category entitled "**absolute no's**" (traits that result in automatic elimination of a candidate). An organization, which approaches an opening using this progression, is, at least, started in the right direction. Where the organization is at the end of a search cannot be guaranteed. However, if an organization understands that it needs to begin by shooting for the bull's eye, understanding that even if it only hit the outer ring of the target, it is better than using a process that may end with missing the target completely.

There is a process!

The process is the critical component for selecting the best candidate. Set up a team to do the screening – the collective assessment is better than that of any one person. The team starts by determining the skills, talents, personal characteristics and the temperament needed in the new individual. If there is an inside candidate who meets all the criteria, he should be called in and offered the position, thereby ending the process – the organization should demonstrate the faith it has in his ability. If there is no viable inside candidate, solicit applications and include the criteria in the advertisement. The team then sets up the screening process and picks the questions. The process ends with two finalists – never with just one!

The selection team

A selection team has three direct responsibilities and an interesting positive "side effect" is usually created. The first responsibility of the team is to determine the characteristics that they are looking for in a new staff member. The second responsibility is the trickiest part, determining how they will find out if the candidates have the desired traits. The third step is conducting the screening. The interesting "side effect" is the outcome on the staff members selected to be on the team. Being chosen to serve on a selection team is a real honor. The staff members inevitably take their responsibility seriously and try to project a positive image of the company. Those designated to be on a selection team also have a sense of ownership of the person who was chosen and will go out of their way to help the person be successful.

Creating a team with diverse perspectives is important for several reasons. The additional people will add different opinions/perspectives. They also will add technical knowledge. Even the best supervisor may not fully understand the responsibilities of a clerk, but a fellow clerk does. Along the same lines, the clerk could tell whether a candidate really had similar responsibilities based on follow-up questions that the supervisor or personnel person may not understand or think to ask. Selection committees also project an image of the company. There is an impact on candidates when they realize that they are being interviewed by future peers, not just a potential supervisor.

There is a caveat on choosing selection team members. This is one place where an idealist can be devastating (see *Idealist versus Pluralist*). An idealist's extreme views can come out during the screening, even if his own questions were not selected as part of the interview. Since an idealist does not want to compromise, he can hold the committee back from making a selection based on his belief about a candidate. Idealists have a place in an organization, but being on a selection team is not it.

The first meeting of the selection team should be a brainstorming session. One member should record all of the traits desired in a new employee given by team members. After all of the qualities are listed, each one is placed by consensus under one of the priorities (required, wanted, etc).

The advertisement

One of the keys to finding good candidates is to advertise where potential candidates are likely to find out about the position. The team also has to be sure the advertisement includes all the characteristics and skills required.

Selection

Below is an example of how the priority list is developed. Since most people are familiar with schools, a school model was used as an illustration. For the list of priorities below, assume that there is an opening for a person to be a third grade teacher in a suburban school district.

- **Required** are those traits that are a minimum, i.e. College degree with State certification.
- **Needed** are those traits that help a person to be success ful in the position, i.e. one year teaching experience or three years working as a substitute teacher.

- **Wanted** are those traits that give better assurance that the person can match the organization's needs, i.e. experience at grade 2, 3, or 4; experience teaching in a similar district, or an advanced college degree,
- **Liked** are those traits that would really help, i.e. experience in grade 3, graduate of a specific college (some programs are better than others)
- **Desired** are those traits that really put the person ahead, i.e. certification as a reading teacher, etc.
- **Bonus**es are those traits that become the sifters. If there are two candidates who are considered equally qualified these traits break the tie, i.e. ability to play an instrument, ability to coach a sport or to lead a club.
- **The absolute no's** are the traits that eliminate candidates. They could be something such as a criminal conviction (depending on the offense) or even something as straight-forward as the ability to work with the other teachers on the team.

Essentials for the Selection Team

An organization needs to establish rules by which selection teams operate. Everyone who is selected or volunteers to be on the team has to agree to the rules. If a person cannot agree to the rules, than he cannot be on the team. It is also important that everyone on the team is a *one* (see *One's hire One's*).

- Team members must be sworn to secrecy. In every organization there are staff members who would like to see a specific person selected for a position. Providing the criteria or, worse yet, the questions would give a candidate an unfair advantage.
- At the "paper screening" it takes two or more people to put forward or to cut a person. No one person has the power to cut on his own. All candidates who do not meet the priorities listed as **required** should be eliminated at this point.
- Members of the selection team must share with the other members any knowledge they have of a given candidate.

There are occasions when a member of the team did not realize that he knew one of the candidates – the most common situation is that a candidate is the member of the same (health) club as the person on the team. When this happens the team member needs to stop the interview and share his discovery. The team member is not eliminated but it will explain why he is behaving differently.

- People who agree to be on the interview team must attend all the interviews or they lose the right to vote – they may still speak on behalf of or against a candidate, but they cannot vote.
- The same requirements have to be used with all of the candidates. One candidate cannot be asked to submit a writing sample unless all are required to submit a sample.
- The same questions must be asked of all the candidates and should be asked by the same team member in order to maintain consistency. Teams may choose to type the questions for the candidates and allow them to answer at their own pace. This method eliminates the possibility of a questioner's vocal inflection influencing the candidate's response.
- Each selection team member, in the interview, needs to be given a copy of the questions and the list of **needs**, **wants**, **likes** and **desires** on which to keep notes. Note: the **required** traits do not appear, because the candidate should not have been granted an interview, if he did not meet the requirements.
- After the last interview and before a discussion, the team members should rank (or score) the candidates on each of the needs, wants, likes and desires. Then one of the team writes all the candidates' interviewed names' across the top of a board or large paper and the criteria down the sides creating a grid. Finally, in each square created, each member of the team scores each of the candidates. This is everyone's personal impression. Silence needs to be maintained to get honest first impressions during the

recording process.

- The next step is the discussion. By looking at the completed grid, members of the team can see how the various candidates were perceived by the other team members. Some candidates will immediately be dropped from consideration, while other will be discussed because team members had divergent views.

It is not uncommon for the chair to reserve veto power. [Author's note: This system was used for more than twenty years, and I never invoked a veto – the system always worked to provide the best candidates.]

By starting with a paper screening the teams will only be interviewing people who are believed to meet the **needs** and hopefully even the **wants** criteria. If the screening was correct the team will select who it likes from people who meet the pre-established **needs** criteria – success.

When the process starts a member of team will usually ask, "What happens if we do not find someone?" He really wanted to add the words "we like." The answer should be "we don't settle." The organization that is willing to start over rather than settling is the organization that will grow. Settling before it is absolutely necessary lowers the organization's standards.

Additional contacts saying "no"

One mistake made frequently by organizations is failing to show the courtesy of a follow up letter notifying the unsuccessful candidates that the position was filled. This failure makes the organization look unprofessional and may keep a good candidate from applying for a future opening.

Never Hire Someone
You cannot Fire

The primary rule for hiring in any organization is "Never hire someone you cannot fire" – the principle is so important that it could be the first five rules.

Who fits into the category of those who should not be hired? Start with those who have emotional connections to a supervisor or the head of the organization. That group includes children, friends, relatives and anyone intimate. Anyone who fits in these categories rarely has a chance at true success because, regardless of his qualifications, the rest of the staff will believe that he was given the position as a favor. Even if the person is fully qualified, he will have difficulty gaining the respect of his peers. The person will also never be a confidant because the staff will fear the person's connections with the supervisor. Whenever he is around, most of the staff will be on guard. The ultimate problem arises if the person has to be dismissed. In this case there is a double problem, first there is the emotional upheaval that is present in any dismissal, and second, dealing with the boss after he feels the ramifications of having someone close to him released.

Better to never hire than to deal with the fire.

The other group to be eliminated from consideration is anyone who tries to use connections to get a position. A candidate tries to use connections by having someone of influence approach members of the selection committee on his behalf. Connections in this context include both those within the organization and those outside. Inside connections are those people who come forward and call attention to the person's resume. These inside connections usually include the head or anyone else who, on the organizational chart, is above the person on the selection committee. When someone 'above' the person on the committee comes forward, the person on the selection committee feels intimidated. Outside connections are people who are not employed by the organization [such as religious figures, elected officials, executives of other organizations] but who have dealings with the organization, i.e. bankers, accountants, lawyers. This rule of eliminating someone who tries to use influence is not true *if the organization asks* any of the above for the names of potential candidates.

Once hired, people who used contacts to get a position will feel that they owe the person(s) who interceded on their behalf. These people begin the position with divided loyalties, part to the organization and part to their "godfather." People who use connections to get a position will use them again if (or more accurately "when") the system puts obstacles in their way. To add to the dilemma, a person who uses connections will usually "bad mouth" the organization to his "godfather" as soon as his needs are not met.

There is one other necessary clarification. If an employee, who will ultimately be a peer of the new employee suggests a candidate, the person may be worth considering. This is a case where the *one's hire one's* theory is in play. Staff members who are considered to be good employees will usually only recommend people whom they feel will help the organization. Note: the use of the word "recommends." Staff members have loyalties to people who are not affiliated with the organization; they may feel a responsibility to put someone's name forward but not want to actually recommend the person. When a name comes from a staff member, take the time to determine how the staff member feels about the person.

How to avoid hiring someone you cannot fire? If the person is not in a protected group, by far the easiest way is to set the screening standards so that they do not fit the criteria. If he is not interviewed, there will not be as much of a feeling of loss.

To people who use connections, getting an interview implies they have the position.

One's Hire One's;
Two's Hire Three's

"A weak link wants to be sure that there is weaker link. In that way, he is confident that he is not the point where the chain will break."

It is human nature to recommend candidates who will make us feel comfortable and make our performance look good. Therefore, members of a selection team will recommend someone who will make their positions easier. In the case of successful people, those who have strong self images (*one's*) will hire others who are strong, with a high potential for being successful (other *one's*). The reason is simple: successful people realize that hiring other successful people will make their jobs easier. Conversely, people who are substandard will recommend people who are even weaker, to be sure that they look good by comparison.

One of the fastest ways for any organization to move forward is through recruiting; therefore, it is essential to find the right people. This requisite sounds simple, but it is tremendously difficult to accomplish. The question here is: will the person responsible for hiring recruit the best person? Simply put, if you don't have the right person making the hiring decision, how can you expect the best candidate to be chosen?

To understand this concept, the levels *one*, *two*, and *three*, need to be defined. *One's* are effective employees; they get the job done with little supervision. *Two's* are satisfactory, but have demonstrated weaknesses; they can do the job but it is a challenge. *Three's* are on the border of satisfactory with tendencies to dip below the line; people who are *three's* need extensive supervision.

Employees who are either a *two* or a *three* struggle every day. It is only natural that these people are intimidated by those who do a good job (*one's*). People who are struggling don't like to have *one's* as competitors; therefore, they will recommend people who are weaker than themselves. Their logic is simple;

they don't want the organization to consider replacing them.

In the case of a supervisor who is worried about his job, or who is in a position where he is in over his head, he does not want people who will challenge him. He is being questioned enough by those for whom he works; he does not want questions from those he supervises. It is the nature of the pecking order. In the depths of a *two's* mind, it is somehow acceptable to be near the bottom, as long as there is someone still lower.

From a recruiting perspective, never expect a weak link to bring in to the organization a strong employee. Supervisors must find a way to recruit around the weak link. It is better to have a supervisor from another department (if the person is a *one*) select a new employee than it is to have the direct supervisor, if that person is a *two*.

Wonder how *two's* get hired? It is when there are no *one's* available.

"Bless the organization that has a 'three' in a hiring position because it is in big trouble."

The Devil You Know
Versus
The Demon You Grow

Whether a supervisory position is filled from inside the organization or by someone from outside, eventually we will see the person's dark side.

Whenever a supervisory position opens, the same question is asked, "Do we go outside (the organization) or can we promote from within?" This question is asked by those who will supervise the new person, and by the people whom that person will supervise. During the ensuing discussions among the supervisors, the names of potential "inside" candidates, especially the informal leaders will be bantered around (unofficially, the staff holds its own caucus). The process of reviewing the qualities of the "inside" candidates takes place before the position is advertised, and the results often determine if an outside search is warranted. If the question of succession is asked as soon as the opening exists, the answer is already clear. The names of those who should be considered for promotion ought to be well known even before an opening occurs.

In their initial conversations, the supervisors will discuss the strengths and weaknesses of each of the potential internal candidates. As with all positives, the strengths will be general in nature and, therefore, limited in number. The candidates' weaknesses will be specific and, therefore, more numerous. This is the period during which the organization examines the "**Devil(s) it knows**".

Established organizations understand that the selected person will have weaknesses. These organizations make the decision to go "outside" when they are sure that the internal candidates' weaknesses exceed their strengths. And that their weaknesses will lead to their failure, or will change the direction of the organization in a negative way (see *Arrows*). The general categories that are examined in considering the inside candidates are experience, knowledge and respect of their peers. Some organizations will even consider position levels (see *Position Levels*) in their considerations of the "**Devil(s) it knows**."

An organization that looks outside to fill a supervisory position is searching for an ideal candidate or "savior." During the screening process, the team will select someone they like, rarely someone they need. During the interviews, the organization will build hopes and expectations based on the responses that were given by the successful candidate.

If the person hired is from "outside" the organization, he will eventually begin to show weaknesses (we are all human). A thorough background check may limit the scope of the Achilles' heel, but weaknesses will be there. One new factor that could not have been predicted, even by the new supervisor's former employer, is the effect of the promotion on the new supervisor's ego. The new person knows, at least subconsciously, that he was the winner, the champion. Naturally, his ego begins to grow. As his ego inflates, he will feel capable of meeting the expectations of his new position; this ego growth can lead to arrogance or to careless mistakes. The alternative is that he is not as capable as anticipated, and he begins to make mistakes; then his ego begins to deflate. In either case, this is the beginning of the "**Demon You Grow**".

In advance of a supervisory opening, organizations should know whether they are going to go with the "Devil they know" or "The demon they grow".

Who to Follow?
The loser, not the Winner

The quickest way for a new supervisor to appear to be successful is to follow a person who was perceived to be a failure.

Conversely, the better the predecessor, the more difficult it is for the successor to appear successful. In too many cases, when a supervisor replaces someone who was successful, the staff has placed that predecessor on a pedestal, even if he did not deserve the accolades.

One of the most important predictors of the success of a new supervisor is the status of the person whom he replaces. Even knowing what is in the following pages, people will take positions in which they will almost surely fail.

Following the Loser

When a supervisor is lucky enough to follow someone whose exit was desired, he is in a position to look superior by almost any comparison. It does not matter why everyone was waiting to see the first supervisor leave; the simple fact is that the worse the predecessor looked, the better the successor appears.

From incompetence to professional burnout, and including personality traits, there are numerous reasons why some supervisors fall into the category of "the vanquished." When the perception of the supervisor deteriorates to this state, his staff hovers like vultures, waiting for him to exit.

In following a "loser" (let's call it what it is), the successor needs to find out why, in general terms, the person failed. The two primary reasons a supervisor is unsuccessful are incompetence or personality.

Incompetence as a supervisor means that the person was not able to resolve problems. If the supervisor was incompetent, then it can reasonably be predicted that the organization is in some degree of disorder; the problems were not resolved. The new supervisor is now faced with a dilemma: the organization knew that his predecessor was unfit but does not know that it is in turmoil (they held the predecessor responsible for all the problems). The sooner the problems are identified and addressed, the more they will appear to be the result of the incompetent supervisor. The longer the old problems are on hold, the more they belong to the new supervisor. Therefore, for his own success, the new supervisor needs to move quickly; establishing which issues need immediate resolution, which can wait, and which will never be resolved requires triage. Utilizing a system which determines those problems that can be treated quickly and will yield the greatest effect is imperative.

If the problem with the predecessor was incompetence, the organization needs to recruit someone who is a catalyst, a leader and a unifier. A supervisor who is a catalyst gets things going; he will change the status quo. The person needs to be unifier because, although the staff may have been united against the predecessor, there is no reason to believe that they were in agreement on the solution to the issues. The downside for the organization is that supervisors who have the characteristics necessary to fill the requirements of catalyst usually stay in a position for only a few years.

Although it is generally good to follow someone who was incompetent, there is a potential pitfall. In many cases, the staff is under the impression that they held the organization together. When the staff feels that they succeeded despite the former supervisor, they have not acknowledged the turmoil within the organization or their partial responsibility for it. The turmoil exists, but responsibility is masked by sweeping the issues under the rug. The successor has to figure out how to sweep the problems out from under the rug without getting a mess on himself or the staff.

If the reason for the predecessor's failure was his personality, the staff will be unified in their distaste. If the person was competent, just repulsive, there may not be any real institutional issues except letting go of the past. In that case, personality is a major factor in choosing the replacement.

In the replacement of a supervisor who was disliked because of his personality, an organization needs a person who is outgoing enough to meet with the staff and sensitive enough to make the staff feel that their feelings are understood and appreciated. Most important, the person has to be a stabilizer, capable of calming the troubled waters. In this case, a unifier is not necessary, since the staff was already unified in their distaste for the first supervisor. In this replacement, the social skills of the supervisor will be one of the key elements to resolving the issues of the department. The person selected for this type of situation should be planning to stay for several years.

Following the Interim

The second best situation for a new supervisor is to follow the interim. Interims are brought in because they are strong as either **leaders** or **managers**. The key talents to be a successful interim are strong skills and definitive personalities. Organizations hire interim **managers** if they feel that the organization is on solid ground; they also hire interim **leaders** if they are trying to resolve overt problems. In either event they have a clear understanding of what is needed in the short-run.

The new supervisor needs to be aware that, if he follows an interim who was a *leader*, the interim was brought in to address specific problems. If the interim was successful, the problems were resolved and will not need to be dealt with by the new supervisor. If the interim was a *leader* and unsuccessful with some problems, the staff is aware of the issues and anticipates that some changes will be made – the stage is set. In either situation, the new supervisor is in a position to succeed.

If a new supervisor follows an interim who was a *manager*, he is in an excellent position because the staff is anticipating change. Whenever a supervisor leaves, the staff, at least at the subconscious level, expects change to occur. The staff may not like or want change, but they are aware that it will happen. If the interim was a *manager*, little change occurred. Now, with the new supervisor, the staff is doubly ready for change, once for the first person who left, and a second time for the interim leaving.

Follow your opposite

The third best position for a new supervisor to hold is to replace someone who is not like himself. Pendulums swing, and so do the requirements for supervisors. If the predecessor was a strong *manager*, the staff will be ready for a *leader*. If the predecessor was a strong *leader*, the staff will be a ready for a *manager*. Supervisors, and those aspiring to be supervisors, need to know themselves and to understand that they will be more successful if they follow someone who was not like themselves.

Following "the lifer"

The fourth best position to be in, as a new supervisor, is to follow someone who was in the same position with the same organization for a very long time. It is easier for a supervisor to survive for a long period as a *manager* than as a *leader*. *Leaders* push people, and eventually the staff tires of being pressed. When the staff has been pushed as far as the leader can get them to move, he either has to leave or stop pushing. A supervisor who stops moving forward starts managing. *Managers* survive longer than *leaders*, because they do not tend to rattle cages. *Managers* move slowly and steadily. Therefore, at the ends of their careers, almost all, if not all, people who stayed in the same supervisory position (*lifers*) are *managers*.

Managers are easier to follow than *leaders*. If a manager follows a *manager*, he should be successful because the staff is use to change taking time. If the new supervisor is a *leader*, the staff will resist slightly, but will anticipate change. In either case, the new person has a reasonable chance to be successful.

The supervisor who follows a lifer inherits a staff that is not accustomed to change. The staff will not necessarily resist change, but change was not a part of their daily processing. When following a *lifer*, it is best to begin by instituting incremental changes that are the ideas of the staff (See *Looking Brilliant*). In time, the staff will be ready for a more rapid pace for change.

Following the Winner

The most difficult way to be viewed as a successful supervisor is to follow a person who appeared to be successful in his position; those people are almost universally believed to be *winners*. Note the cautious words "appeared to be successful." In reality, the new supervisor will discover that there were dents in his predecessor's armor. Even with those dents, chances are that the organization has successful systems in place, thus making it difficult for the new supervisor to make his own mark.

There are three possible scenarios when following a *winner*. First, the predecessor was considered to be a *winner* by the staff, but not by his supervisors. Second, he was considered to be a *winner* by his supervisors, but not by his staff. The final option, he was considered to be a *winner* by both his staff and his supervisors.

The biggest problem lies in following a person whom the staff thought was successful, but whose supervisors had a different assessment. In this case, those above the new supervisor will be expecting changes, and those whom he supervises will feel that change is unwarranted. The supervisor who finds himself in this quagmire will need a delicate approach when making changes.

Following a person whom the staff held in lower regard than the person's supervisors is the easier situation. In this case, the staff wants change, and the supervisors are not requiring it. Unless the new supervisor is being micro-managed, change desired by the staff, as long as it is not too extensive, can easily be implemented without the supervisor's supervisors even knowing that it is happening.

Following a person who was considered by everyone to be successful can only be accomplished by someone who is a manager. If everyone perceived the first person to be successful, no one wants change or believes that change is necessary. Therefore, those involved in the selection may talk leadership, but they really want the status quo.

Following the Institution

There is a situation that is impossible. No one can successfully follow a person who has attained the level of being considered the *institution*. People who are considered to be the *institution* are rare, but there are still some supervisors who have held their positions for so long and have demonstrated such a high level of success that they are considered by all to be the *institution*. These people are not part of the organization; they are its heart, soul and brains. These people often started the organization. They were both a leader and a manager, and they consistently made correct decisions. The problem is that even these people were human, and there is dirt under the rug. Depending upon the duration of the person's tenure, the dirt may be inches deep. The problem now exists that it is impossible for the new supervisor to get the dirt out without getting most of it on himself.

When an organization needs to replace someone who has attained the level of being the *institution*, it should look for an interim. The sad reality is that the supervisor who follows an *institution* will be an interim, even if he was not hired for that role.

There is a difference between looking good and being good. Being good will last an entire career, while looking good is like beauty – it fades with time.

It's Only Business
The Art of Firing

A negative evaluation, a reprimand, or the termination of a person's employment, are three actions that are almost always perceived backward. The recipient of any of these three actions should think of the action as "just business," while the person making the decisions should think of the choice as very personal. In reality, we do exactly the opposite.

In supervision, there are many difficult decisions; none is more emotionally wrenching than those which mean the end of a person's employment. Taking the position that the decision "is only business" helps the individual who makes the decision; while to the person receiving the news, it is as personal as anything gets.

In the movie The Godfather, when the rival families are going to the "mattresses," one advisor turns to Michael Corleone and says, of the shooting of his father, "It's not personal; it's only business." The concept is simple; hard-times and hard-decisions are necessary to weed out the unfit. Put slightly differently, decisions are right when they are based on the greater good of the organization. Thus, to be a successful and fair supervisor in difficult times, it is often essential to make a decision based on what is best for the organization, even if it hurts an individual. This "organization first" attitude is especially important when the decision impacts the future of another person. Truly great supervisors, however, understand the effect of those choices on individuals, and they add a human element to the decision.

If a supervisor thinks too personally, the potential exists that he will become apprehensive and, therefore, ineffective. The line between what is good for one person and what is good for the organization is usually very clear. Making the decision is almost never tidy because it impacts someone who is known to the supervisor. To reduce the strain on their consciences, most supervisors try to depersonalize a dismissal.

One extreme example of the personal factor is that, in some ways, it is actually easier to close an entire operation than to dismiss an individual. In a mass layoff, the impact is on a group, not a person; thus the personal element is limited. It is also easier to let someone go because of his lack of seniority rather than his performance, because seniority can be determined clearly, while performance may be more obscure.

Supervisors are often divided by how they deal with the personal factor. On one hand, personalization takes away from objectivity. Supervisors who allow the personal element to control too many decisions let their organizations fail from the outside because they lose competitiveness. On the other hand, supervisors who are too objective let their organizations fail from the inside because they lack recognition for individual differences, and the staff loses loyalty. Finding a middle ground, whereby a supervisor can still relate on the personal level, while maintaining a fair level of objectivity, will prove to be invaluable to an organization. Organizations with supervisors who possess these unique skills not only survive but thrive as well.

There is one really strong downside to the concept of "it's not personal"- when it happens to you. When it is you who loses a job or receives a negative evaluation, it is as personal as it gets.

When those who hired you are gone
Watch out you will follow

*There is a career expectancy for a supervisor that is meas-
ured in large part by the turnover of others.*

Throughout the period a supervisor holds a position,
there are a series of transitions in the opinions that people have
regarding his performance. Almost always, these transitions
eventually take a downhill turn. Since the greatest expectations
are placed on people at the top of the organizational chart, this
change in perceptions is especially true the higher the supervisor's position ranks within the organization. This shift in opinions occurs more quickly, if a person is brought in from another
organization, rather than having been promoted from within.

The first transition occurs when a newly-appointed
supervisor stops being regarded as the champion or hero
brought in to solve all problems. At that point, he is no longer
seen as the savior, but instead is seen simply as a supervisor – a
part of the institution. This change in perception takes place
with both the person's supervisors and his employees. When the
transition from champion to supervisor is complete, he is considered to be part of the system – the system that he was hired
to correct.

If the supervisor stays long enough, a second transition
occurs. He is perceived to evolve into a member of the "old
guard." Being part of the "old guard" is different from being part
of the system. Being part of the system allows for the possibility
of change – the old guard, in contrast, resists change and is,
therefore, an obstacle to new vision or real growth.

*The process of this transition is governed by the succession
of those who hired the supervisor and those whom he hired.* To
those who hired the supervisor, he is the future. As those who
hired him leave, their replacements do not remember the reasons for which he was hired or the problems that he resolved.
The new people see the supervisor as part of the institution
which they joined. To the new people, he is the present. To these

new people, he is not the savior, but rather part of the reason for the current problems. When the change in those who supervise him reaches a *critical mass,* the supervisor will need to seek new employment.

For a supervisor, a similar transition in perception happens with those whom he supervises. Those who were on staff when he was hired, and who agreed with the need for his skills, will support him (Group 1). To those who were on staff and did not support him - the "old guard" (Group 2) – he is just another cog in the system. They were there when he was hired, and they expect to be there after he is gone. Those whom he hired (Group 3) have the most influence on the cycle. Those who were hired very quickly after he started, the initial wave (Group 3a) see themselves as a part of the answer. They are "his team," and they will support him even longer and deeper than those who saw the need for his skills (Group 1). Those who were hired after the initial wave (Group 3b) see him as the part of the system. As problems arise, this group (Group 3b) will see him as the source of the problems – after all he is a part of the system.

Clearly, an organization changes and revitalizes itself through recruiting. With the exception of the departure of members of the "old guard," all replacements shorten a supervisor's tenure. A supervisor who keeps himself technically current, optimistic, enthusiastic and contributing can retard the process, but the transitions still happen.

There is one major exception to this process – those who are able to stay long enough to be considered "the institution." They are not part of the system; they are "the institution." These rare individuals are allowed to operate on a very different scale. They have evolved from administrating policy to being policy.

Everyone new is good for the organization and bad for you.

Why did you get the position?
To solve a set of problems

Whether he was promoted from within or brought in from outside the organization, a supervisor needs to recognize that he was selected based on a belief that he could solve a set of problems. Perceptions of his success will be determined by how quickly and how accurately he identifies those problems and starts to implement solutions. With some problems, the new supervisor will be successful, with others, he will pacify the issue, and there will be a group of problems that he will never resolve.

Even though no one gets them all, there are still ways to be a star. Someone who hits over .300 in the majors or scores the winning touchdown, as time expires on the clock will be a star – at least for the moment. In short, a supervisor can be a star by getting most of them right, or he can be a star by merely making the big decisions correctly.

*Supervisors are **superstars** when they are both consistent and make the big score.*

A Different Way to Look at Candidates
How will the person affect the social structure?

People believe that the successful candidate for each position in an organization was chosen based on how well his skills and experiences were believed to "fit" a given position.

The ability to determine where the person will fit in the pecking order of both the formal and informal structures greatly reduces problems during both orientation and periods of acclimation. A new staff member automatically alters the organization's "pecking order." Understanding where the new person will "fit" in the social structure provides an opportunity for real change in the infrastructure (Informal leadership). For example, if an organization is having a problem with one of the informal leader's and a major component of the person's source of power is his athletic prowess, bringing in a better athlete will erode the informal leader's standing. The new person, just by his presence, will change the informal leader's status. When an organization considers a candidate's "fit" in the pecking order, it has a chance to create systemic change.

Pick people who make a difference – change the social order

Evidence of a Supervisor Who Lost
Who gets to hold the door?

There is a sure way to gauge whether the former supervisor lost all his support. Did the staff hold a lottery, and the winner got to hold the door when he left? And the runner-up got to hold the door for his replacement?

Evidence that a Supervisor was a Winner
The party was full

The best evidence that the former supervisor was a winner is that his goodbye party had to be held in a ballroom which was filled to capacity.

Dress to Fill
A position

Employment services advise candidates about what clothes to wear for an interview. Usually they recommend that it be the best clothes that the person would wear to work in the position for which they have applied. When considering a candidate's outfit at an interview, the supervisor needs to evaluate the appropriateness of the candidate's choice.

Very few people ever lost a job because they came to the interview over-dressed, though many people missed opportunities because they came to the interview too informally attired for the position. Painful as it may seem, the best advice for a candidate, applying for a supervisory position, is to dress as his father would expect because the person who conducts the interview will probably be his father's age.

Dress to Interview
Watch the news

When interviewing for a supervisory position, there is a simple barometer for appropriate dress. Wear the same styles as those who are reporting the national news. For men, that means the suits, shirts, and ties. The same is true for women, including fashionable colors, appropriate necklines and hair style.

Words to listen for
Examples of successful experiences
During an interview, listen for the words, "I always try to…" versus "I have done…." The latter is much stronger and shows success, while the former indicates a lack of commitment or self confidence.

Another behavior to listen for
Why is the person available?
During most interviews candidates are asked why they are leaving their current position. If the applicant's response cites a reason that is a complaint about his current supervisor, he is giving a clear warning about his candidacy. People who complain about one boss will usually complain about the next. During the interview explore his reasons (probe); determine the nature of his concerns. There could be acceptable reasons. After all, there are supervisors who do not deserve their positions. In general, if you hire a person who cited his current supervisor as the reason he was leaving his last position, he will cite his next supervisor as the reason when it is time to move on.

The counter thought is simple, has anyone ever lost a position exclusively because he did not denigrate his former boss? Probably not, since interviews are supposed to be positive situations and every candidate should be as positive as possible.

A candidate's attitude about his work environment will only go down hill after the interview, so why start halfway down the hill?

Never hire or promote a friend
It makes you both look bad
There are lines where friendships end; and true friends never expect someone to cross that line. The employment or promotion of a friend makes both parties look bad since the friendship, not the person's qualifications, is always seen as the

reason for his selection.

Trying to maintain an employee/employer relationship that includes a friendship is like trying to navigate a minefield, either a tremendous amount of time is spent on avoiding the potential outburst or it is just a matter of time until the explosion detonates. The relationship also places a blindfold on the person as he enters the minefield. Every staff member knows if his supervisor has a friend on the staff. That friend is suddenly treated like an extension of the supervisor, and vital information is cut off (the blindfold). *Informal leaders* will use the friend as a source of information, and as a way to plant information (some of which may not be true). Worse yet, the friend often becomes a scapegoat. In any event, the friend will never be fully trusted by the staff.

The promotion of a friend does not work even if the friend is selected to succeed the supervisor when he leaves. The friend, even with the former supervisor gone, will live under the cloud of having secured the position through their friendship, not his qualifications. The staff has the same feelings and concerns about the "friend" even if the original supervisor is gone. Far too often, a successful supervisor's legacy is tarnished by the appointment of a friend to fill his position.

A supervisor can recommend a friend to work in another department. The key is that someone else hired the friend based on the friend's qualifications.

Knowing who can be promoted
An organization can never be too prepared!

There is a true story of an incident in which a supervisor was taken to the hospital at noon. The problem was heart related, and it was immediately understood that the person would not return for at least a month, and probably even longer. Within two hours, four temporary promotions were enacted by those in power. They selected an assistant from another department to fill in for the supervisor. An entry level supervisor, from yet another department, was moved into the role vacated by the assistant. A staff member was moved into the entry level supervisory position. The same day a new person was hired, from a pool of candidates from a previous search, to fill the staff position.

The original supervisor never came back to work. All of the positions were eventually posted and every person named in the first two hours got the permanent position. The reason was simple; the name of every person involved in the process had been considered long before an opening occurred.

Interviews Have Two Sides
Do they want you? Do you want them?

During an interview candidates become so focused on the questions being asked and the challenge of getting the position, that they miss the second major issue. Do they want to work in the organization? If examined carefully, interviews provide the candidate with a better picture of the organization than the organization gets of the candidate.

Studies have shown that interviews are one of the least valuable sources for determining whether a candidate will be successful on the job. This is logical, since interviews place a candidate in front of strangers who hold the key to his future. Unless this is the nature of the position, the interview is not measuring performance on the job. What the interview does provide is a snapshot, however blurred, of the person's personality, knowledge and background. The interview also provides the candidate with a picture of the organization. Candidates should ask themselves how did the organization present itself? Were the members of the interview team positive about the organization? Did they have a potential peer on the selection team? Is there evidence of trust?

Supervisors should look at they way interviews are being conducted by their organizations. The examination should begin with the initial phone call setting the date and time through walking the candidate to his car. What is the image that the organization projects? There may be a lot of candidates, but there are never enough quality candidates. The supervisor wants to set a tone so the best candidates want to work for his organization.

When Does an Interview Start?
Sooner than anyone thinks.

An interview is usually considered to be the time spent in meeting and questioning a candidate. Supervisors, however, should consider redefining an interview as the entire time involved in the process of screening candidates for a position. This would begin with reviewing the applications and end with walking the candidate out of the office after the final meeting. By this definition, the interview starts with the examination of the person's written submissions. Reviewing a pile of applications or resumes is time spent on getting to know the candidates.

There is one other step in the process that unquestionably counts, and that is the phone call setting up the interview. After the interview is scheduled, if the person who set it up says something like, "When I called, he was really upbeat and sounded interested." It is a *first impression* to build on. If instead, the same person says, "He wanted to know about salary and benefits." That is a *first impression* to overcome. Any supervisor who does not inquire into the impression of the people who made the initial contact with a candidate is missing important information.

At the very least, the interview starts the minute the person walks on to the organization's property. From that moment on, anything he says, does or implies counts.

When does the Interview End?
When a candidate is offered the position.

At the very earliest, the interview ends when the candidate leaves the organization's property – not when he leaves the interview room. Candidates who think that the interview ends when they leave the interview room are misinformed.

One organization, when considering candidates for supervisory positions, would tell the finalist that, after the final meeting, they would be taken to lunch by a senior level supervisor. Like many organizations, this one had a practice of having finalists meet with several interview groups. The lunch was presented as a way to unwind after the stress of the interview. Although the candidate's host for lunch was from another department, he would be glad to answer any questions. The organization used this practice for years, yet it never appeared that the candidates realized that their "host" was a full part of the interview team. The host's role was to examine the candidate's behavior when he thought he was in an informal setting.

A Few Years of Excellence
or a lifetime of mediocrity

There is one frustration that happens too often when the interviews end. Those on the selection team discuss who they feel is the best candidates for supervisor. If one person has clearly demonstrated the skills and capacity to perform the responsibilities a member of the team will inevitably say, "If we hire him, he will use this position as a stepping stone and will leave in a few years."

The implication is clear, the person who made the remark would prefer a long period of mediocrity to a shorter period of true leadership. There are two questions that are presented by the one remark.

Is mediocrity really better for the organization than excellence? The answer to this question has to be answered by the organization.

The other question involves what the person would have to do in order to use the position as a stepping stone? The answer is clear. To move up the candidate will have to increase productivity in a creative and harmonious environment. It is hard to find a long list of what is wrong with the effect a person who is excellent will have on an organization.

If in doubt, hire the brightest
He will only let you down in a big way.

Even with the best criteria, the best team, and the best process, it is not always clear which of two candidates would be better in a position. There is a one final criteria that almost always works when the field is narrowed to two. Hire the brighter of the candidates.

Regardless of the work that a person is expected to perform, the greater the ability of that person to think of options and alternatives, the more likely it is that he will be able to come up with systems to perform the task more efficiently. Efficiency leads to productivity, and productivity leads to success.

There is a disadvantage to hiring the brightest. If the work does not challenge him sufficiently, he will develop other ways to use his intelligence. The way that can be either good or bad is when the person becomes a practical joker (the effect depends on the jokes). If he is bored, the other option is the person makes careless mistakes. The really bad outcome for a bored, bright person is when he comes up with ways – sometimes illegal – to use his brain.

The problem with an early start
Who can help a career?

The only people who can help in climbing the career ladder are those on a higher step.

No one can help a person to get a supervisory position higher than the one they have themselves. This makes the youngest, and first of a group of peers to start to rise in his career, face the tallest hurdles. All his associates are below him on the ladder and he has no one to help him on his ascent

No one wants the caboose gone.
It gives the supervisor time

There is a bond that naturally exists among staff and against their supervisors. Even if a person clearly needed to be dismissed, the staff, as a group, will rarely admit it to the supervisor. The best a supervisor can hope for is that one or two of the employees will acknowledge their agreement with his decision in a private setting.

Communications

"Do you have any idea what happened in that meeting?"

The Movement of Information

Communications
The one thing that goes both right and wrong.

It is not what was said, as much as how it was heard.
It's as much in the messenger as it is in the message.

People communicate in numerous ways; however, exchanges fall into two categories: the messages which the person intended to send and the messages which the other person actually received. For a variety of reasons, the sending and the receiving are rarely the same. It is this dichotomy that creates the impression that communications are often mishandled.

Communications include everything from body language to voice intonation, from meetings to memos, from emails to personal visits, from formal presentations in an auditorium to casual greetings in the hall. Whenever two or more people meet, greet, or write, they are conveying a message. Supervisors may wish that communications were as simple as the choice of vocabulary, but as words are being said the listener is looking at the expression on the speaker's face and hearing the intonation of his voice. Even the choice of a written message over a spoken one has its own implications.

People spend years learning how to write and to speak; yet, it is through listening, watching and experiencing that people take in most new information. Although there are people who seem to instinctively understand body language and voice inflection, the interpretation of these less formal ways of imparting information is rarely taught formally. It is through the understanding of the informal messages, that one grasps what people are really thinking.

Supervisors need to be very aware of the signals that they are sending because problems develop when the goal is to transmit one message and the style of the presenter provides a very different meaning.

Vocabulary
Words Reveal More Than the Sounds They Make

Some expressions and automatic responses tell much more about a supervisor than he may want people to know.

A supervisor's best words: *"You're doing a great job."* Everyone likes to feel important, successful and that he is making a contribution. Critical to job satisfaction is praise from both fellow workers and supervisors. Supervisors contribute to satisfaction by providing a sense of appreciation; there is even an expression that "praise is worth more than a raise." If a supervisor only says, *"great job"*, the person feels valued, but the impact is quickly lost. If, however, he uses the words, *"great job* on (the name of a specific activity)", the person has something identifiable to hold on to. Making the compliment specific provides the employee with recognition for an identifiable task or service and shows that the supervisor recognized the person's contribution in a specific situation. The more specific the compliment, the longer the impact will last. Compliments can be repeated if they are sincere; after all, they don't wear out.

A supervisor's easiest word: *"Yes."* Saying, "Yes" to a request may appear, on first glance to be easy, and such an affirmative response allows the action or plan to proceed. However, in too many situations a supervisor says, "Yes" without proper consideration of the request. If given too quickly, his, "Yes" frequently comes back to haunt him when difficulties arise later. Worse yet the quick, "Yes" can establish a precedent, and everyone remembers a precedent, especially if it is self-serving. A supervisor may respond with the quick, "Yes" thinking that he is performing a "good deed." But, at some point he will learn there is truth to the old adage "No good deed goes unpunished." "Yes," may be an easy response, but it often has disastrous consequences. Like dealing with anger, it is best if a supervisor counts to ten before saying, "Yes."

A supervisor's strongest word: *"No."* There is nothing wrong with the word, "No." It might be nice to always be in a position to say, "Yes" but in reality, there are many times when an affirmative answer is not appropriate. "No" when said arbitrarily or vindictively is inappropriate; however, "No" said with a brief explanation of the reasons for the negative response provides the employee with two choices. He can either accept the negative response, or he can modify his idea to address the reasons given for the rejection. Either action by the employee provides him with closure.

A supervisor's worst word: *"Maybe."* If a person is told, "No" he can rethink or drop the idea. If a person is told "Yes" he can begin to implement the plan. The word, "Maybe" although occasionally necessary, leaves the employee hanging with no sense of direction. Worse yet, it gives the supervisor the appearance of either being weak, lacking authority or being indecisive. "Maybe" should only be used when additional thought is required. When confronted with a specific question to which "Maybe" is the appropriate answer, the supervisor needs to make a final decision as soon as possible.

A supervisor's weakest words: *"Let me check."* If a supervisor says that he needs to "check," he implies that he either lacks authority to give an answer or that he does not possess knowledge that is basic to his position. For example, if an employee asks, "Can I take a vacation day on (date)?" and the answer is, "I need to check the contract," the supervisor has shown that he is not knowledgeable about a contract provision for which he is responsible. A better response is: "We are only allowed to let (number) people take a given day. Let me see how many are out already." Then the supervisor needs to investigate and get back to the person. "Let me check" is a losing line.

A supervisor's most important word: *– a person's name.* A name is unique; it belongs to a person. In effect he owns it, and it is specific to him. When a supervisor uses a person's name to address him, the employee gains a sense of identity and recognition. Caution: avoid cute names or nicknames; these are often derogatory. This warning is true even if the person's peers use his nickname.

A supervisor's most defensive words: "Let me explain." As soon as a supervisor has to explain an action or, more likely, the failure to take action, he has made a mistake. The reasoning may have appeared to be correct when the decision was made; however, the very fact that it has to be explained means that the decision did not turn out well. It is almost impossible to use the words, "let me explain" without appearing to be defensive. The best advice is to tell the truth and put an end to the issue.

A supervisor's most challenging words: "How can we do this?" The word "We" immediately implies a team effort/approach and shows respect for those involved in the discussion. The words, "How can ..." imply a belief that, through collective wisdom, a solution will be found and that everyone's ideas will be considered in resolving the issue. The words challenge each member to become involved in a total team effort.

A supervisor's most optimistic words: "We can__." "We can __", unlike the challenging words, "How can we__?" shows a fundamental belief in the individual(s) involved in the decision. The words project a feeling of support, trust and success. A supervisor with an optimistic perspective radiates the confidence necessary to accomplish the task. It is true that success can only be achieved when there is a belief that the project can be successful.

A supervisor's most pessimistic words: "We can't __." The words, by their very tone, radiate an acceptance of defeat. "We can't __" is much worse than "I can't ___." When a supervisor uses, "I can't ___ " he is showing a lack of belief in himself. Employees do not like to see this self-directed pessimism, but it is worse when they are under the impression that the supervisor does not believe in them either, as evidenced by the pronoun "we" in "We can't."

A supervisor's most thoughtful words: *"How is this different from, __?"* In just five words, the supervisor has demonstrated several different skills. By referring to a parallel situation, he has shown that he remembers a precedent. Through this parallel, he has also implied the answer to his own question; whatever the solution to the first situation should be the solution in this situation. He has allowed the person the freedom to explain a difference or to accept the answer. A supervisor with a good memory has an advantage because of his awareness of precedents. All supervisors can be assured that those whom they supervise will have an excellent collective memory of precedents.

A supervisor's most empathetic words: *"How can I help?"* There are times when every individual needs support. Identifying these times, which are often not verbalized, is a skill required to be a successful supervisor. The problems may stem from work, home, a technical issue or interpersonal relationships. No matter what the supervisor's perception of the weight of the issue, to that person effected his tribulations are real. The supervisor's offer of support needs to be sincere. On more than one occasion, a supervisor has learned that one of his employees or an employee's family member has a serious health issue. By lightening the load of that employee, to the extent possible, the supervisor will gain respect from the person and the person's peers (*chip theory*).

A Supervisor's words for fixing blame: *"And the problem is?"* These words automatically either establish the systematic reason for something not working or, if the problem is personnel, the words make the person to whom they are asked begin to establish responsibility. If the problem is technical or in the system, the words allow the employee the ability to demonstrate the problem. To be a truly successful supervisor, one must be able to determine when a problem that appears to be in the systems is really in personnel.

A Supervisor's most foolish words: *"I never thought of that."* These words imply a lack of foresight on the part of the supervisor. Instead of projecting confidence, these words reflect his limitations. The same message, totally redirected is conveyed by the words, "good idea," without a name. By depersonalizing the response, the supervisor eliminates implications of his own lack of creativity.

A Supervisor's wisest word: "Why?" There are many ways in which the word "why," spoken in a non-threatening way, indicates a supervisor's wisdom. It buys time for the supervisor to reflect, since the person who was asked "why," will need to explain his position – which takes time. Since the best way for the employee to show "why" is to cite precedents, the supervisor is showing respect for standards. "Why" also shows that the supervisor expects people to be able to explain themselves – a sign of his respect for the staff. Finally, to answer the question "why?" requires reflection. This reflection usually draws the correct answer from the person.

A supervisor's most dreaded words: "I have to let you go." Supervisors don't want to hear these words or to be in a position where they have to say them. Despite what some people think, it is emotionally draining for the supervisor when an employee has to be dismissed. This is especially true if the supervisor tried to help the person improve his performance through fair and accurate evaluations. By providing the employee with training and support, the supervisor has taken a vested interest in the person. The failure of the employee's commitment to work has an emotional impact. If the supervisor did not support the person, and is now letting him go, then the supervisor is inevitably feeling a degree of quilt. In any event, the words hurt.

A Supervisor's most dangerous words: *"I don't understand."* These simple words appear to be non-threatening but they are rarely as simple as they seem. These words call on an employee to justify an action. The best aspect of the words is that, if they are said openly, they allow a legitimate answer – there could be a reason for the action. If, however, there is not a reasonable answer, causing an employee to explain his action is even more powerful than the supervisor making an accusation. Example: "I don't understand why you took an extra 10 minutes at break time" is dangerous to an employee because it requires justification. Whereas, "You took an extra 10 minutes at break." Is a statement that does not require the employee to justify his action.

The most unpredictable words for a supervisor to hear: "Oh, by the way, ___" (sometimes replaced with "Have you got a minute?") There is no way that the supervisor knows what is to follow. There may not be a way to control the use of these words; however, these words always preceded the most difficult problems and some of the best news.

Everyone's most appreciated words: *"Thank you."* Say them often, they are free!

Body Language
Screaming in Silence
Or
Actions really do speak louder than words

It isn't what people say that counts, as much as what their bodies tell.

Without a doubt, one of the essential skills of a successful supervisor is understanding body language. The way a person sits, his stature when standing, his lip and eye movements, are clear indicators that speak more honestly than the words the person articulates.

Body language has dialects, just like spoken language. The major conveyers of body language are: shoulders, angles of the head, glances of the eyes, use of hand gestures, movements of the mouth, posture, including leaning, and inflection of the voice. There are cultural components that play into each of these elements. For example, people from the Mediterranean Basin tend to use hand gestures much more frequently and intensely than people from northern Europe or Asia.

As an illustration of body language: a person who is being truthful will tend to keep his shoulders square to the person who is questioning him; he will maintain eye contact and speak with limited hesitation. However, if a person is trying to avoid answering a question, he will tend to glance down, to hesitate before speaking, turn his shoulders at an angle, and try to put something, even his crossed arms, between himself and the person who asked the question. Instinctively, people who are not going to be truthful will almost always try to cover their mouths. This is best illustrated by the person who rubs his nose while answering.

The language of the body is too complex to be examined in depth in this book; however, it is important to recognize some of the traits. For those who understand its relevance, numerous books and articles are available which illustrate and explain aspects of body behaviors.

Understanding body language is especially important for a supervisor who is preparing to conduct an interview. Increasingly, consultants are being used to prepare résumés and to coach candidates on how to answer standard interview questions. Many of these same services offer classes on how candidates should portray themselves during an interview. In effect, the supervisor is looking at an actor, not the real person. However, the automatic responses of a person's body will ultimately tell who the person is and how he feels.

Some people can naturally read a person's body language; for the rest of us in supervision, it is a set of skills that are essential to master. It is not enough to read a book on body language. Examples of body language that are mutually observed need to be discussed and ideas tested.

The language of the body is more honest than the language of verbal expression.

Save a Tree – Gain Respect

Those who need a memo, never read one.

One of the most ineffective things that a supervisor can do is to send a memo to everyone when it only applies to one or two people. Example: one or two people are beginning to make a practice of coming to work late. Action: Send a memo telling everyone to be on time. Reaction: Those who are usually on time resent the memo, and those who are late will not even bother to read it. Outcome: No long term change in behavior and the supervisor is seen as too weak to address those with whom he has a concern.

There is nothing wrong with expecting a person to meet reasonable standards or addressing him directly when he doesn't!

First Impressions
Something to build on or to overcome

Everyone is constantly making a first impression.
There are two different types of *first impressions.* The obvious *first impression* is the one made when people meet for the first time; this is an initial or traditional *first impression.* Less obvious *first impressions* are formed whenever there is an initial contact between people. *First impressions* can be just minutes apart, as evidenced by the comment, "What happened? I just saw him, and he was in such a good mood."

The time span varies, based on which study is referenced, but data indicates that people have begun to make assessments of a person within a range of just seconds to three minutes of contact. A positive *first impression* is something to build on, while a negative one is something that needs to be overcome.

The major components of both an initial impression and a *first impression* are dress, personal grooming, the person's posture, facial expression, body language, and voice. Whenever a supervisor walks into a room, those present immediately try to assess his mood. The supervisor's persona projects a feeling of how the meeting or the day will unfold.

Dress: People instantly judge what others are wearing. Judgments are made based on style, cleanliness and quality. Another section of this book is devoted to dress; however, within the context of this section, examining the dress code of an organization provides an important backdrop for a first impression. As a meeting begins, those in attendance examine the person who called the meeting; they are looking for a suit jacket, a loosened tie, anything that projects a feeling or tone. The way in which a person is dressed provides either a negative or a positive *first impression,* in about equal measure.

Grooming: Because people do not want to be perceived as elitist, they rarely talk about grooming. This does not mean that grooming is not evaluated. Grooming may appear to be a personal choice; however, today people work in such close proximity that good personal hygiene is a must, not an option. Strong smells are distracting. This is true of natural odors or manufactured scents. Cleanliness is linked very strongly to promotions. Since we live in a first world society, poor grooming will most certainly yield a negative first impression.

Facial Expressions: Nature teaches us to read facial expressions. Fear, joy, frustration, anger and appreciation are all clearly visible in the expressions on a person's face. People who smile a lot are generally accepted more quickly than their counterparts who smile less. People may be nervous when they first meet, but a smile at the initial meeting will lend itself to a positive first impression.

Body Language: The way in which a person carries himself, his posture, style and speed of walking, and grace, are all factors in the first impressions that he gives. A supervisor walking quickly and deliberately down the hall gives a very different impression from the same person leaning against the side of a cubby, facing into the work space, chatting with the person inside. The elements of body language, especially a handshake, are key components to a first impression.

Voice: One underrated element of a first impression is the person's voice. With experience, one can detect nervousness, humor, and self-confidence just by listening to a person's voice. It is said that people can tell whether a person is smiling by hearing his voice on the radio.

First impressions are given as well as received. Supervisors need to acknowledge and evaluate the impression that they give as well as the impressions they form. This is done through honest self-analysis. The simple question is: "What would someone think if he saw me right now?"

With the wrong first impression, there may never be a chance for a second impression.

It's a Cover Up
What You Wear

Clothes may not make the man, but they may determine where he is headed.

Consciously or unconsciously, everyone matches what he wears to his plans for the day. When selecting clothes in the morning, an anticipated visit by representatives from the corporate headquarters suggests one outfit, while a golf league outing after work would suggest something quite different.

Justly or unjustly, as a culture, people are expected to act in accordance with the clothes that they wear. This cultural norm of clothes matching behavior results in every organization having a standard of dress. The standard of dress may be formal or informal but it is present. Even if not spoken, organizations are assessed by the dress of the employees.

All levels of society know and expect dress codes. There are uniforms designed to instill confidence, such as those worn by pilots, police, and military personnel. There are also the less formal, yet equally distinguishable, uniforms worn by those on a surgical staff. Law offices tend to be very formal, with professionals wearing dark suits. There are even "uniforms" we expect of those who work in construction or in retail. People dress in uniforms even for leisure activities, whether playing golf, biking or swimming. Who hasn't heard the expression, "Did you see what he has on?" The meaning of the expression is clear; his outfit did not fit the situation.

When interviewing for a position, a candidate needs to determine if he fits the message that his dress is providing. In seeking a position one should always wear the best outfit that he would wear to the position for which he is applying.

Adolescents have one of the strongest senses of dress codes. When a leading female rock star wears a new style in a video, within one week, a sizeable portion of girls from age 12 to 14 are pleading with their parents for a similar outfit. Another example is the youth who consider themselves to be Goth.

127

These adolescents project themselves as having rejected the standards of society, yet, because of their dress code, they are among the easiest teens to identify.

Knowing that dress codes set a standard, it is reasonable to assess people, **in part**, based on what they wear. People who dress in the center of the dress code of the organization are sending a message of fitting in, while those on the perimeter, whether it is more formal or more relaxed, are sending another message. Usually, people who want to be promoted will dress on the upper (better) end, while those who view work, as a necessary evil, will dress on the lower end of the scale.

There are cautions against the overuse of dress as an assessment tool. People on the lower end of the economic ladder, or those who are just starting out, may only have enough money for one set of clothes which they have to wear at multiple levels. There are also some subcultures who view dress very differently from "Corporate America."

One of the best examples of a person who realized the importance of dress involved a woman who applied for a professional position in an organization that had both professional and non-professional positions. The woman was working for the organization as a non-professional. Her interview for the new position was held during the day. In the morning, she arrived appropriately dressed for her non-professional job. When it was time for her interview, she wore a business suit. She had brought the suit with her that morning and had changed in the ladies room before the interview. Everyone in the interview knew her from her non-professional position, but was very impressed by the wisdom of her choice.

In an interview setting, candidates will be judged to a large part by what they wear. The same is true, even if "political correctness" prevents people from admitting that this is the case, in work and social settings.

What Time Is It?
People think in different time periods.

Effective supervisors know that different employees relate in different time periods. They use this knowledge to talk to each person in terms that he understands.

Whenever a discussion is taking place, it seems that some people just don't "get it." Although intelligent and knowledgeable, they fail to see the reasoning. There is a joke that the people who are not following the discussion are listening in a different time zone. Not true, they were actually thinking in a different time period from the one in which the speaker was talking.

People associate in three different time periods: past, present and future. Unconsciously, most people think primarily in one period, with a backup to a second, and some time spent in the third. Rarely does a person function equally in all three time periods. It should be noted that age has nothing to do with the time period people relate in! These periods are easily identified by listening to informal conversations in the break room. Some people will be talking about what they did last night, while others are talking about their current assignments and others are talking about the upcoming weekend.

When thinking in the past, people draw on experiences. Hopefully, these experiences have contributed to the person's wisdom, thus keeping the organization from repeating past mistakes. This part of a person's mind tells him not to try something a second time, because it did not work the first time. People who relate heavily to the past have limited credit card debt, because they don't tend to buy. On the down side, people who focus on the past are often heard to say, "We never did it this way before." or "We've always done it this way."

When thinking in the present, people are looking for immediate outcomes. They are constantly thinking about what is appropriate now. People focused in the present tend to enjoy what is happening around them. These people want tax relief now, even if it means a greater debt for their children. On the downside, people based in the present tend to have large credit card balances, with many expenses for entertainment or other consumables.

People who focus on the future can be either optimists or pessimists. In either case, "future" thinkers are savers. They have the 401 and 403 plans and are picking out retirement communities while in their forties. Future planners try to avoid debt, but the credit card balances they have are mostly for the purchases of items (investments). These people are constantly planning.

In general, it is no better to think in one of these time periods than any of the others (Caveat: we all believe that the time period we operate out of is the best). Supervisors, who try to talk about the future with someone who is focused in the past, will fail to make their point unless, during the conversation, they acknowledge that changes made in the past helped the organization to become more efficient.

Watching a skilled presenter, it becomes evident that he covers all three time periods with every idea. He presents the history of the problem, follows by what is happening now and then discusses what he believes will be necessary in the future.

For example: When I first came to the company, a very busy switchboard operator was sorting calls. He did his best to transfer people to the appropriate offices. Now calls are handled by an automated answering machine which gives callers a menu of choices. The problem is that when customers are transferred to the wrong office, they are unhappy and feel that the company has become impersonal. In the future, we will add the option of speaking to a customer service representative to the initial menu.

As a supervisor, it is not what time frame you are thinking in that matters; what matters is being able to acknowledge the time frame the other person is functioning in.

Meetings
The Communications Base of Organizations
The Nemesis of Employees

If there is going to be a meeting, there needs to be an agenda, and everyone who attends should have a reason for being present.

A universal perception exists that there are too many meetings, that the meetings take too long and that little is accomplished as a result of the meetings. Said succinctly, meetings can be a waste of time.

Meetings, in any organization, should be held for the purpose of discussions or major announcements. Planned and structured properly, meetings can be productive and cost efficient. Unfortunately, too many meetings are held simply because they are scheduled, then they ramble on too long, and eventually deteriorate into social events or gripe sessions. If there is no need for discussion, or if the discussion will not alter the outcome of the announcement, supervisors may prefer canceling the meeting and sending a memo. There is, of course, the distinct possibility that those who should read the memo are the very ones who will throw it away.

The premise of this section is that meetings should be for the purpose of discussion. That discussion can include an option whereby employees are allowed to vent, even if their opinions will have no impact. Meetings for which the outcome is predetermined may, very appropriately, seem like a show. In organizations, some degree of show is a reality; therefore, some meetings are a place for show.

Meetings are usually considered to be instances in which two or more people come together to discuss topics of mutual interest. Although not always acknowledged as such, meetings take place whenever two or more people have direct contact. Passing someone in the hall may seem like chance, but in reality each person formed a *first impression*, and a meeting occurred.

Although two people casually passing in a hallway is a form of meeting, it is not a meeting in the context of this section, since there was no discussion.

Presentations have multiple formats, so they are also discussed in the *Staff Development* section. Presentations are when a speaker(s) is brought in to instruct the group on a specific topic. Seminars are one of the most common forms of presentation. Seminars are often set up through associations with representatives of multiple organizations present.

Presentations to a single organization, however, should be based on the needs or goals of the organization. The goals under consideration can be short-term or long-term and staff development on the topic should be included. When the supervisor is responsible for a presentation in his organization or department, he needs to be sure that it is properly prepared. Preparation includes giving consideration to the time of day, and the length, format, and environmental issues (i.e. space, refreshments, and presentation equipment required). It is especially important that the supervisor make the presenter aware of any issues. To avoid disappointment there needs to be a clear understanding in place with the presenter as to what topics will be addressed.

Cost: Meetings cost money; this aspect of meetings is often overlooked. The greatest expense for a meeting is often not the direct expenditures for materials, space, refreshments, or transportation. The greatest expenditure is the indirect cost of time spent away from the attendee's responsibilities. For any meeting, the direct expenditures are reasonably easy to determine. The cost, in terms of time, is not as easy to assess. In addition to the actual time spent in the meeting, in some cases the expenditure of time spent on preparing and for travel to the meeting should be included in the calculation of cost. With hourly workers, the cost of time can be determined with relative accuracy because it is the length of the meeting multiplied by the hourly rate plus the cost of roll ups (benefits). With salaried staff the cost is more difficult to ascertain. It would appear to be easy to calculate the cost by computing the person's daily rate multiplied by the percentage of the day that the person spent in the meeting. For example, if there were eight supervisors who attended a meeting that took two hours, and the average salary was $1,000 a week or $200 a day. That implies that each supervisor is paid about $20 per hour, and the meeting cost $320. The problem with this simple formula is that there was also lost supervision, so productivity may have dropped thus driving the cost even higher. At the same time, an idea may have been presented at the meeting that would increase productivity and offset all the cost. The bottom line is that meetings consume resources.

Politics: Within an organization, the concept that *Everything Matters* is acutely evidenced in meetings. Who attends, why they were chosen, where they sit and even the purpose of the meeting can be political. When it comes to the politics of a meeting, the real question to consider is; what is _not_ political about a meeting?

There are meetings, both with an individual or with a group, where the discussion will have no impact but the message needs to be conveyed face to face. Group meetings that require the politics of a face to face interaction include instances in which the organization wants everyone to hear an announcement at the same time. Group meetings include announcements of expansions, mergers and layoffs. In the case of a political meeting with an individual, the organization wants, or needs, the person to hear the news directly. Cases in which a political meeting is held with an individual include dismissals and promotions. These meetings require direct contact but the discussion should be limited. Meetings fitting these descriptions have to be considered political, since the purpose was not discussion, but to convey the image that the organization or the person holding the meeting faces problems directly.

Another form of potentially political meetings are those between supervisors and representatives of the staff (union). In addition to being productive, these meetings often occur just for show, since they demonstrate that direct contact exists between representatives of the parties.

It is important to note that there are political implications in who is not included in a meeting. Because the people who are not invited to a meeting are sensitive to the reasons for their omission, supervisors need to exercise care and use a set standard when scheduling meetings that will be attended by only part of the staff. Supervisors should remember that it is not the people who are invited to a meeting who care as much about who attends as the people who were excluded.

The test of a supervisor's understanding and control of the politics of a meeting is demonstrated in who he calls on and in the order in which people are allowed to speak. Tone, direction and outcome of a meeting are, all too often, set by the sequence of speakers. The selection of the first speaker is similar in nature to a *first impression*; they are something to build on or something to overcome. If people who oppose an idea are allowed to speak first or to dominate the discussion, a reasonable idea may be defeated.

The best way to learn to chair a meeting is by watching others. Learn from their mistakes and successes, remembering that order of the agenda and the sequence in which people speak in the discussion are essential components of the outcome.

Agenda - the prescription for success: Since meetings exist for the purpose of discussion, to make the dialogue as open and thoughtful as possible, everyone involved in the meeting should know in advance the topics that will be covered – an agenda. The agenda should be one page or less.

If the meeting has both a limited attendance (two or three people) and a limited number of topics (one to three), the initial message via email or even a phone call, which set up the meeting, may be sufficient to serve as an agenda.

Throughout this section there are references to "controlling a meeting." The first and most important method of control is determining what will be discussed and how long that discussion will last, (i.e., the agenda). There needs to be a clear procedure for adding items to the agenda, and a firm deadline for submission of items to be considered for the agenda (i.e. all items are due 48 hours before the meeting). If someone wants to add an item to the agenda after the deadline, he must call before the meeting and discuss what he wants added and why he did not submit the information in a timely fashion. Allowing items to be added "at will" allows those who attend to avoid planning, even worse, it allows topics that can be a distraction to consume valuable time. A practice that allows items to be added to the agenda at the meeting is strongly discouraged – another person's failure to plan is not the supervisor's problem.

The agenda should be arranged so that the most important topics are at the beginning of the meeting and the less significant items are at the bottom of the list. The last thing that should happen is for time to run out before the major topics are fully explored. A second advantage to having the most important items first is that it strongly encourages people to be on time. If the agenda includes the amount of time allocated for each item, it allows people who may only be invited for a limited number of topics to know when to arrive and approximately how long they will need to be in the meeting.

If there is to be a presentation by someone who will not attend the entire meeting, it is recommended that the presentation be either the first or last item on the agenda. If the presentation is first, it allows people to set up in advance of the meeting, but there will be a distraction when they leave. A presentation at the end means that there is the distraction of the setup but not a problem as they leave. If the presentation is in the middle it creates a distraction both at set up and departure. To demonstrate interest and support, the chair of the meeting needs to sit through any presentations. The natural impression is that if the chair is not in attendance the topic of the presentation is not important.

Another issue of control is the ability to return to an item that was previously discussed. The ability to return depends on the situation. One way employees evaluate their supervisors is in terms of fairness and consistency in his decisions as whether to return to a topic. If the reason an employee wants to return to a topic is due to his own tardiness, his request should probably be denied. If the reason a person requests to return to a topic was that a later discussion triggered a different potential resolution, it may be practical to return to the item. It is in the best interests of the supervisor to prohibit a practice of returning to an item, unless the person making the request explains to everyone why he feels that there is a reason to revisit the topic.

Often agendas have a section for "roundtable" "good of the order" or some other term that opens the meeting. Regardless of the name, it means that anyone can add anything and, therefore, the supervisor will lose control. To retain control, supervisors are strongly encouraged to have rules for what this open-ended item means and how it is to be used. Better yet, eliminate this section from the agenda altogether and require that the issues that would have come under this item be placed on the agenda in advance. Keeping control does not dictate whether or not people are allowed to raise an issue, rather, how and when they raise the issue.

One way to be sure that the meeting ends appropriately is to have the person who is running the meeting quickly review each topic, summarize the discussion and review who will need to follow up on each item. It is strongly suggested that a written summary also follow. If there was a reason to meet, there is a reason to remind everyone of the outcomes.

Suggestion: If an agenda and support documents require multiple pages, do them in different colors. That way the chair can see a glance what each person is looking at.

Schedule: Meetings fall into four general categories. Some meetings are regular – the same group of people meet according to fixed calendar (i.e. supervisors meet at 9:00 a.m. on the second and fourth Monday of the month). Meetings can be irregular – called when there are sufficient items for an agenda or the items have sufficient weight to require discussion. There are also special or emergency meetings which are called irregularly and have a limited agenda. The worst and best meetings are those that are spontaneous. Spontaneous meetings are those that, at least on the surface, appear to just happen or to happen with very little notice.

In regularly scheduled meetings the scheduling issues include, the day of the week, time and frequency. Arguments for any particular day of the week can have equal value depending on the amount of travel or special projects that historically happen on any given day. The time of day is more significant. Meetings that begin the day set an atmosphere not just for the meeting, but for the day. When a meeting is held early in the morning, those in attendance are less distracted by a problem that happened before the meeting. Meetings that begin the day also have the ability to run over if necessary – the problem is that the start of the meetings may be subject to traffic. Meetings at the end of the day may allow running over, but resentment may be felt by someone who has a personal obligation. Meetings in the middle of the day usually require at least one of those in attendance to leave a project unfinished.

It is usually best to hold irregular meetings at the beginning or end of the day. Those who will attend do not have them on their calendars, and it is usually easier to clear the beginning or the end of the day than the middle. If the meeting is for discussion of a "hot" item, naturally the beginning of the day allows sufficient time for input. If the meeting is more for an announcement or a "pseudo" emergency, the end of the day is better since it allows the people to leave and at least stops the rumors for that day.

The meeting at which an employee is being discharged is always best at the end of *his* day.

"Spontaneous" meetings either just happen or, more frequently, require extensive timing on the part of one of the parties. If a supervisor has regular habits, he can be reasonably sure that if someone "accidentally" meets him in the break room, and begins a conversation with, "Oh by the way," it was not a coincidence. The same format, of being trapped, can happen in the supervisor's office. The best way for a supervisor to control a "spontaneous" meeting is to use *walk and talk*.

There are occasions when it is important to talk to one employee about a very specific issue. One of the best examples occurs when a supervisor suspects that an employee is having personal problems. In these cases, the supervisor may not want the employee to become concerned over a formal meeting, opting instead to just drop by the person's work space and starting a conversation.

Length of a meeting is the <u>minimum</u> time required to discuss the topics on the agenda. The length of time on a topic should be the <u>minimum</u> required to reach a resolution. People in attendance should be allowed to discuss a topic for as long as what they say adds to the discussion. Meetings are often dragged on by those who voice their support or worse yet, give examples of what another person has already stated. For the supervisor in charge of a meeting, it is imperative to maintain control. One suggestion is to say, "Does anyone have anything to add to the discussion that has not already been brought up?" Placing the estimated times next to each item on the agenda will help control the discussion. The downside of establishing a time limit for each topic is that some people are reticent to express their views in a group setting. Looking for body language can reveal to the chair of the meeting that these people have something to contribute, and they should be recognized. Meetings are held to gather the opinions of all those in attendance not just the ones of those who can talk the loudest, longest or most.

How a meeting starts sets the tone for the meeting. There was a senior supervisor who believed that if he scheduled a meeting and someone was late, the reason must be that the person was required to do something more important than meet with him. This was a high standard. Since the supervisor scheduled meetings much earlier than his predecessor, some of the staff resisted, and began making a practice of using excuses for being late. Understanding the importance of maintaining control, he told the group, as a whole, that he expected meetings to start on time with everyone present. In a couple of cases, the practice of being late continued, so he spoke individually to those who were late. One person continued. This person was warned that if he was late again his next request for a conference would be denied. The person persisted and his request for a conference in Las Vegas was ripped up. The lesson was so clear to everyone that attended this supervisor's meetings that a person could be run over if he were near the door to the conference room in the minutes before the scheduled start.

Supervisors who make a practice of starting meetings at the exact time scheduled will soon find that employees will become more punctual. More than anyone else, the person calling the meeting needs to set the standard by being on time.

The Site of the meeting and how the space is arranged do a lot to set the tone. There are three general considerations when selecting the site: Which space will be appropriate for the meeting? Is it the supervisor's space? The employee's space? Or a neutral space? There are also general parameters for the best arrangement of the site that should be considered. Key elements are the arrangement of the seating, accommodations for writing, and the shape of the table(s). The supervisor needs to know the impression that he is trying to make, then use the space to send the appropriate message.

The supervisor's space – when and how to use it. Any place where the supervisor has primary control should be considered his space. Spaces generally associated with the supervisor are his office and any conference room, especially if the conference room is in the area of his office. By holding a meeting in "his" space, the supervisor has asserted control. Likewise, by bringing the person to his space, the supervisor places himself in a position of authority. This use of space to demonstrate authority is even stronger when the meeting is in his office rather than the conference room.

Within his space, the supervisor has a multitude of subtle messages that he can send based on the nature of the meeting. Should the supervisor decide that the appropriate site is in his space, the next question becomes the seating arrangement.

If the supervisor sits behind his desk, he holds more authority. One can equate the "behind the desk" position to that of a judge on his bench. If the supervisor moves to a table, he creates a more even playing field. Sitting in a chair on the same side of the desk as the employee levels the playing field even more.

A supervisor, who has to deal with a difficult employee, will want to sit behind his desk. This is contrasted to a supervisor electing to sit on the same side of his desk with a person who has had long and loyal service and has chosen to retire.

The supervisor's conference room has nearly the same impact as his office except that it is less formal. Because conference rooms are controlled by a table the supervisor can send different messages based on where each person sits. If the supervisor sits at the head of the table, he is in the authoritative seat. If the supervisor sits along the side, he has lowered the barriers. A round table is less imposing than one that is rectangular. Another advantage of a round table is that the supervisor can clearly see everyone's face. Observing body language is much easier at a round table, since none of the participants can lean back out of the supervisor's view.

In regular meetings, where each person sits sometimes becomes an issue. Given a choice, people tend to always sit in the same seat. If the supervisor notices that the arrangement has become a problem (side comments or inappropriate behavior between certain attendees) he may want to put people's names on the agendas and place those agendas where he wants the people to sit. Subtle messages usually work.

The staff's space – when and how to use it. Getting out of the office and demonstrating approachability is a universal trait of good supervisors. This means that he goes to his employees. If the issue that the supervisor wants to discuss is not disciplinary, it is usually best to go to the employee. In his own space, the employee has a greater comfort level because he is sitting behind his own desk, thus providing him with a greater feeling of authority. When a meeting is in the employee's space, intimidation is lower, and a more open discussion will ensue. If the supervisor stands while meeting with the employee he allows himself to escape more easily. The problem is that, while standing, there is projection of dominance. If the supervisor sits, it increases the employee's comfort level, since the employee is not looking up at the supervisor. Sitting also implies that the supervisor expects the discussion to take some time, and the employee will probably be more prone to extend the conversation.

When (note the word "when" not "if") a supervisor needs to see an employee about something in the employee's personal life that appears to be impacting his work, the best thing for a supervisor to do is to go to the employee and to make the setting as comfortable as possible. These meetings are essential when the employee or a member of his family is experiencing health problems, is moving, is having financial issues or some other transition. When the employee's mind is consumed by personal problems, he needs to know where he stands in his professional life.

The neutral space – when and how to use it. Neutral spaces are places not in the control of either the supervisor or the employee. These spaces are usually outside the organization's building. A meeting in the neutral site allows the two to meet without one having to come to the other (sometimes considered a humbling move). Neutral spaces include places like restaurants, the break room (sometimes the employees think this is part of their space) or when walking outdoors. Neutral spaces are those in which neither party has the physical advantage. Additionally, no one has a desk to use as a shield. If a meeting is held in a restaurant, the tables almost always make the parties even because they are either square or round; there is no end to the table. Neutral spaces are sometimes the best sites for resolving issues. One of the best times to use a neutral site is when problems exist between peers, but a resolution is possible, through an open dialog in a setting where neither party appears to have the advantage.

It is important that the room used for any meeting be set

up in advance. Having the space open in advance of the meeting allows for the socialization of the group being called together. Having the space prepared is especially important for those who have to travel to the meeting; a reserved room will provide a place for them to congregate. It also allows those who will be attending to have some informal dialog (team building). The practice of having the room available in advance also eliminates the potential excuse for people who come late.

Attendees at a meeting are the group of people who fulfill one of the following criteria: they were either required to be present, should be present, or were an invited guest. Careful consideration needs to be given to who will be asked to attend a meeting. Since meetings incur significant cost, both politically and financially, those invited need to be limited according to an established criteria. At the same time, meetings can be a learning experience, so people may be invited who are not expected to participate but may gain from the experience. Some of those in attendance will be there because, if they were not invited, they would create a political backlash that is not worth the cost of the seat.

The issue of whether a meeting is voluntary is often raised. Since all meetings are intended for discussion, why would the supervisor not want everyone he invited to attend? Unless a meeting is solely for an announcement, and therefore no discussion is necessary, why would it be voluntary?

When the agenda is set, it should note if the meeting will be cancelled because someone, who is key to the discussion, cannot be present.

One of clearest political messages a supervisor can send is in naming the person who will chair the meeting if he is called out.

Control is an essential for any meeting to be productive. The best control is not apparent, but just seems to happen. Here are some suggestions to avoid problems:

- Have an agenda and follow it – eliminate the ability to add items while the meeting is underway.
- Have a practice of starting and ending on time.

- Have a practice that those in attendance only talk when called upon.
- Have people who will contribute to the idea speak before calling on the detractors - order matters. Taken one step further, after a detractor talks, challenge any points that may cause the meeting to digress.
- Avoid private debates – allow each person to talk only once on each topic – an exception may be made if the person is directly impacted by the topic.
- Be consistent in the application of the policies of running the meeting.
- Summarize all meetings. The summary demonstrates that the supervisor listened and understood the discussion. The best way to show that input was important is to draw from comments made by those in the meeting. When giving examples, the supervisor needs to be sure to put the person's name to his comment.

The question always follows: "What should be done if someone does not follow the designated procedures?" The answer is straightforward. Supervisors must have a progressive plan and apply it fairly. An example of a progressive plan is:

1. Be sure that the procedures for meetings are understood by discussing them at a meeting – in other words, one time make the procedures an agenda item.

2. Be sure that the procedures are summarized and sent to each person in writing – evidence that they all knew what was expected.

3. Talk alone to any person who chooses not to follow the procedures.

4. Require everyone to take a different seat at each meeting. This will stop the side exchanges. If a person continues, talk individually to the offender.

5. When summarizing the meeting, leave out contributions made by the offender – Nothing makes a verbally dominant person feel rejected faster than to have it appear as if he had not spoken.

6. Warn the person a second time that, if he continues

to ignore the practices, he will not be invited to attend.

7. Do not invite to a meeting a person who will distract from the process.

Suggestions for holding more successful meetings.

1. Only meet when there are things that <u>need</u> to be discussed.

2. If possible, take any meetings with an individual to his space.

3. For group meetings, have an agenda, procedures and a summary of the meeting.

4. Set a professional standard by starting the meeting on time.

5. Plan in advance the order in which people will be allowed to speak. The outcome is often driven by the order of the discussion.

6. Debrief all meetings. All supervisors have a confidant. After all group meetings, sit with the confidant and get his input on what went right and where improvements can be made.

7. Have the confidant sit across from the supervisor. By sitting across from the supervisor, the confidant is in a better position to observe body language of people sitting to the sides of the supervisor.

8. Be critical. After any meeting consider what could have been done to make the meeting go better.

Summary

The principle elements of any meeting are: agenda, schedule, length, site and attendance. The underlying issue is who is in control. There are also the secondary considerations of politics, format and cost.

Final Thought

It may seem that there is a lot of planning necessary for a meeting. When a supervisor starts, this is indeed true. With time and experience, good practices will develop. The issue is not how much time went into preparing for a meeting; the challenge is to hold a successful meeting.

The Spectrums
Like a cowboy, we all have a place on the range.

As communication improves, so does performance; therefore, supervisors need to use every vehicle available to reach their staff.

There are a series of spectrums involving one's emotions, personality, thought processing and political perspectives that are important for each supervisor to understand. These spectrums provide insight into the background for how each person interprets words that are read or heard or even events that are witnessed. A person's place on each spectrum also effects his performance, socialization and his natural aptitude toward a set of tasks. These spectrums are like filters, which place values on input.

If a supervisor understands these spectrums and recognizes where, on each spectrum, each member of his staff lies, he enhances his ability to communicate with each individual. By respecting how each employee interprets input, the supervisor improves the chances that the person will actually listen to him.

Because an individual's personality impacts his ability to complete certain tasks, an understanding of the various spectrums allows a supervisor to develop more realistic expectations for each member of his staff.

The first premise that applies to spectrums is that everyone fits somewhere on each spectrum. The position a person occupies will vary based on the environment – people are often different in social, home and work settings.

The second premise is that anywhere a person falls on each continuum is okay. However, an individual's personality may hold him back in his position for various reasons. Note: it is a human trait to believe that where *we* are is the best place (in this case, on each spectrum). The truth is that organizations work best with diversity along the spectrums.

The third premise holds that understanding these spectrums helps in recruiting the right person for a position.

The fourth premise is that with an understanding of

these spectrums and where each person falls on each, a supervisor is better equipped to form effective teams of staff members.

The issues that these spectrums help to address are how to:

- Communicate more effectively
- Develop reasonable expectations for others
- Build mutual respect

There are *eleven* different spectrums mentioned in this section, with several discussed in detail. The preceding section on *Listening in Different Time Zones* could have been placed in this section, but was given its own section because it was so important to effective interactions, yet is so often overlooked.

The spectrums

The first four spectrums come from Carl Jung's research on personality types. This is just an overview to remind those already trained of their importance and to encourage others to become trained and to understand how personality matters both in the selection of staff and in supervision. For those unfamiliar with the theory, there have been volumes written on these four personality types, and there are workshops devoted to understanding and using the concepts.

Sensing and Intuition are generally considered to be the way people take in information. People have five senses which serve as the pathways for absorbing input. People who are considered as *sensing* gather their data by the five senses. These people prefer what can be seen, heard, felt, smelled or tasted. They trust that which can be measured and documented. These people tend to function very much in the present, relying heavily on their own experiences to determine what is occurring at any given moment. *Sensors* are great employees because they get the job done with self-directed follow through.

In contrast, people who are considered as *intuitive* gather information in the same way but are more interested in what the information means: they are looking for associations and possible ways of using the information. *Intuitive* people believe in their intuition and are looking for meaning in all things. People who are *intuitive* focus heavily on implications and

146

inferences for the future. They trust their imagination, inspirations and hunches. People who are *intuitive* are trying to change things; they live in the future. While *sensors* are better at noticing things and remembering facts, *intuitives* look at situations with an eye for determining what it means and the consequences of an action. Because they see associations, people who are *intuitive* are quick to see the solution to a problem. With *intuitive* people, the reasoning used for a decision is less important than the outcome. The downside of *intuitive* staff members is that they become bored too easily and too often work from the cuff.

Supervisors need to talk through ideas with *sensors* step by step emphasizing why any change is important to the present. For *sensors* the reasoning used for a suggestion is just as important as the decision itself.

Supervisors need to plant ideas with people who are *intuitive* and then walk away, giving them time to digest the effect. When planting the idea, the supervisor needs to remember to focus on the needs of the future.

The spectrum **Judging – Perceiving** addresses the way people live their lives. In some ways, the categories could be considered by the more common words structured or spontaneous. These traits indicate the importance of closure on an issue. *Judgers* enjoy structure; they seek to regulate and to control. Note: *judgers* are not necessarily judgmental of others or of issues - they like issues resolved. People who are *judgers* are great employees because they have a great work ethic which includes being happiest when the job is complete.

In contrast, *perceivers* are spontaneous and are happiest when their lives are flexible. *Perceivers* like to understand life, not necessarily to control it. *Perceivers* are not particularly perceptive; in fact they may have a problem seeing things accurately. *Perceivers,* in this context, like to have options. A *perceiver's* greatest enjoyment is in starting a new project or task. The biggest problem with *perceivers* in the workplace is that they are often procrastinators.

If a supervisor wants to lead or to change the direction for an organization, *perceivers* are essential, since, by their very nature, they will allow for flexibility. If a supervisor wants to manage an institution, (improve outcomes through established systems), *judgers* are preferable as employees, since they like order and systematic processes.

The way that people change their minds is in the third area of personality type known as **introvert – extrovert**. People often relate these words to outgoing versus shy. In simplest terms *introverts* focus their energy inside themselves, i.e. they typically suppress enthusiasm. They gain their strength from being alone. *Introverts* are difficult to get to know since they tend to be private. *Introverts* may talk or joke a lot but they do not expose what they are thinking or feeling through their humor. *Introverts* need to think through a problem for themselves. *Introverts* need time alone each day to process what is happening

In contrast, *extroverts* need the company of others. They get charged when they are out with others; i.e. enthusiasm seeps from their every gesture. *Extroverts* are easy to read. It is almost as if they typify the expression "what you see is what you get." Not surprisingly, *extroverts* change their minds through discussion or debate.

Supervisors need to be mindful that *introverts* need to think through a problem; they rarely are convinced of the worthiness of a new idea during a meeting. *Introverts* will accept or reject an idea after thinking it through by themselves. Regardless of the amount of pressure exerted, *introverts* take time to get on board. They like memos, since having the idea in writing provides them with time to reflect.

Supervisors need to remember that *extroverts* need to talk through a problem. They will not respond well to a memo, since it is information that is merely presented without discussion. *Extroverts* take more direct time, but are on board with a decision quickly.

Thinking - Feeling deals with the way a person makes a decision. This area deals with how logical and impersonal someone is as compared to someone who is more empathic. To a *thinker*, a decision is correct if it is logical, even if it may not be in the best interest of a fellow employee. In contrast, a *feeler* places a greater value on harmony within the organization.

Thinkers tend to take a step back in an effort to be impersonal in their analysis. They place a very strong value on logic, justice and fairness. To *thinkers*, there should be one standard for everyone in the organization. It is therefore only natural that *thinkers* see flaws in emotional decisions. *Thinkers* are critical of any decision that lacks logic. Because *thinkers* place a high value on candor they are often described (by *feelers*) as heartless, insensitive and uncaring. Since *thinkers* are logical, they tend to be motivated by a desire to achieve.

People who are *feelers* immediately try to assess the effects that problems or actions have on others. *Feelers* are sensitive to their environments causing them to be empathic, seeking harmony. While *thinkers* want only one rule, without exceptions, *feelers* see the exceptions to the rule. Their desire to please others is more important than the desire to please themselves. They are appreciative of the efforts and feelings of those with whom they are in contact. Their sensitivity causes the *thinkers* to describe them as overemotional, irrational and vulnerable. A *feeler's* desire for harmony causes him to seek a tactful solution. They accept that people can have feelings on an issue regardless of whether those feelings are logical. *Feelers* are motivated by an appreciation of a person's own worth.

Supervisors need to show the logic and fairness of a position to the *thinkers* while showing the *feelers* how it will add harmony.

As mentioned before, this is was only a very brief overview of personality types. These four spectrums are useful in trying to determine the suitability of a person for a position. It also helps a supervisor think of ways to present the problem or solution so that the person will accept the outcome.

The study of Political Science (always sounded like an oxymoron) provides definitions for **sacred** and **secular** in the context of discussions for organizations. These terms are usually used in regards to a traditional religious definition; however the principles also apply to how individuals perceive an organization. *Sacred* in this context considers the organization alone (micro analysis) while *secular* sees the organization within the context of other organizations (macro analysis). In organizational terms, someone who is *sacred* sees issues and answers from within the organization. Employees with a strong *sacred* belief hold that all promotions should be internal, since an outsider could never truly understand the unique needs of their organization. People who hold *sacred* values resist change, especially if they believe that the change is being imposed by forces outside the organization. In a negative way, *sacred* value can be summarized by the expression, "We never did it this way before." On the positive side, *sacred* values provide a sense of pride in what an organization has accomplished.

Secular people, because of their macro perspective, embrace change as necessary for any organization to stay competitive. A *secular* person starts by looking at the competition or other outside forces. Although *secular* people will not necessarily resist an internal promotion, they will want to be assured that the person promoted sees issues other than just the internal problems.

When addressing any issue, a supervisor needs to examine how it was addressed in the past (*sacred*), what change is being proposed, and the effect on the future (*secular*). In order to gain the support of those on the staff with *sacred* values, change has to be presented as addressing an internal need. *Secular* people need to understand how the change will allow the organization to remain competitive. When it comes to change, *sacred* people have to be pushed and *secular* people need to be contained.

Staff members also fall into a range regarding whether decisions should be based on what is best for an **individual** or for the more general benefit of **society** (the organization). People who are driven by the needs of the *individual* believe that society builds off each person, while people who are inclined toward *society* perspective believe the individual gains when *society* adheres to high standards. In a situation in which a person violated an organization's policies, people who start from the *individual* mindset will want to know the person's reasons for the violation. They will also want the consequences to be based on the offender's personal needs. People who are *society* based expect that the outcome will reinforce the standards of the organization.

Every staff consists of people who have leanings that are considered to be **optimistic** and those who are more **pessimistic**. The upbeat, get-it-done attitudes of *optimists are* beneficial to any organization; however, their outlook can cause them to miss potential obstacles. Although they can appear to be "a drag," *pessimists* are also important. The reasons why they think that something might not work allow the supervisor to prepare for contingencies (the devil's advocate).

Idealists believe the mission of the organization is achieved when a set of values is in place. **Pluralists** see the mission of the organization as achieving a consensus from its employees while knowing that the compromise is rarely ideal. Idealists are dangerous in an organization because they do not want to compromise on even one of their standards. Pluralists are dangerous because they will compromise on their principles even when they should hold firm. Virtually all successful leaders of organizations are pluralists. (See also *Idealists steer us Pluralists lead us*)

On first glance, the spectrum of **generalist – specialist** could be mistaken for a pluralist – idealist. A *generalist* is similar to a pluralist while a *specialist* could be confused with an idealist. The difference becomes apparent when examining the difference between a *specialist* and an idealist. In one case the classification is based on knowledge while in the other the categorization is based on a belief. A *specialist* has a deep, although possibly narrow knowledge of some topic, while an idealist has a deep although possibly narrow belief in some value. *Specialists* are usually most successful in staff positions, while generalists tend to be more successful in line positions. *Generalists* because they are broad thinkers, set up processes (how to increase sales) while the *specialists* that work with them set up the specific systems (inventory controls).

It would be incorrect to assume that a **unifier** always works in the best interests of the organization, as it would be inappropriate to assume that a **diversifier** is not necessary in an organization. Although a *unifier* has the skills and personality to bring people together, there are times when people need to work independently in order to perform their tasks efficiently. *Unifiers* are essential for building a team attitude, while *diversifiers* are better at assigning tasks to individuals. Supervisors, with a large staff, are usually selected because they are *unifiers,* bringing people together for a common goal. These same supervisors are most successful when they have, as their assistant, a *diversifier*, a person who knows how to assign tasks to others.

One of the skills that alternates between people who head an organization is to be a **catalyst** or a **stabilizer**. When an organization is in a period of rapid growth or needs significant change, it selects as a head someone who is a *catalyst*. These people have mastered the ability to push in such a way that a proposed change is accepted. Following a period of accelerated growth, comes a period of unification and stabilization. The organization now needs to be headed by a *stabilizer.* On the staff there are people who push for change (*catalysts*) and people who want a managed structure (*stabilizers*). Supervisors need to control the pressure of the *catalyst* while capitalizing on the base created by the *stabilizers* on the staff.

There are people who are very **systematic**; they like to have tasks that are sequential. There are other people who can handle anything as it comes along (**random**). In technical enterprises, *systematic* people are essential; while in broad-based organizations with multiple facets and diverse goals people who adjust rapidly (*random*) are key. *Systematic* people expect things to be explained in order, while *random* people expect the explanation to be direct and to the point.

There are other spectrums that have not been formally

studied but that supervisors may consider significant. Among these other spectrums are: ambition, health, intelligence, outside activities and the desire for goals. In each of these unstudied spectrums, the question remains the same; "how does a supervisor need to address the person so that they will understand the message?"

Each of these spectrums acts like a lens. They add a degree of perspective on the words that were used. In order to be understood, a supervisor needs to respect the location of their staff members on the spectrums, explaining decisions in terms that the person will understand and with respect for that person's individual values.

Filling the circle

When an organization looks for what it needs in a new staff member, it should take time to look at what it already has. If the current staff has a strong leaning toward the end of any of the spectrums, the organizations should look for balance by finding someone to offset them on the other end.

Opposites Attract
In equal degrees

The old expression that "opposites attract" is probably most important in the concept of spectrums. In this context "opposite" is controlled by how far someone is from the center of the specific spectrum. The closer someone is to one of the ends, the more he needs someone close to the opposite end to provide balance. Someone who is near the center on a given spectrum needs someone the same distance from the center on the opposite side. If balance cannot be found in one person, it can be averaged over several people who, combined, are the same distance from the center as the initial person.

Imagine what it would be like to have to supervise a staff that was all pluralistic. With no idealist to counterbalance the pluralist, agreements could be reached quickly but without meaning.

Round Pegs
in round holes

Supervisors need to look at the characteristics of people who are successful in a position then look for new people who have similar characteristics. People are successful for a reason.

How People Judge Others
and themselves

When it comes to the term "judging," in the more general sense there is an excellent line that can guide all supervisors: *"We judge ourselves by our intentions; we judge others by their actions."*

Keeping Current
The benefit of email

Over the last ten years, the use of email for communications within organizations has radically changed the ability of a supervisor to keep current. As cell phones took away the ability of supervisors to truly get away, email has placed organizations in a real time structure. The advantage of email is that supervisors can stay current with their organization even while out of the office. Unfortunately, supervisors are too often expected to keep up even while they are on vacation. The balance is determining how to use email without being used by it.

The Problem with email
What are the protocols?

Like with all radical changes in the way organizations operate, the development of the technology of email is far ahead of the development of its protocols. With email, a supervisor can make the mistake of believing that he can keep abreast of the issues in his organization from inside his office instead of being out working with the staff. Successful supervisors are visible and approachable. Whether it is intentional or accidental, email has facilitated a supervisor's ability to disappear into his office.

There are a lot of problems with the use of email; one of the most serious is its brevity. Supervisors need to realize that, when they send an email, they cannot see the recipients facial response. The message sent by email may not have been intended to offend, but if, in fact, the person who received it was offended there is no way for the supervisor to know from the person's body language. Worse yet, if the recipient questions his interpretation, he may have others read the email in order to get their impressions. Suddenly, the supervisor is being judged by what he wrote.

Here are a few suggestions to help with the use of email:
- *Never ask a question by email that can be asked in person.*
- *Use email as a way to get out of the office, not a way to stay in it.*
- *When receiving praise everyone likes to see his supervisor's face; something that is impossible when the praise is sent via email.*
- *Remember: if the employee perceives that there is an issue, there is probably a record being kept of the emails!*

One solid guideline for emails from a supervisor to a member of his staff is- *if there are more than ten words required and it is not strictly information, go see the person.*

Important Footnote on the word: "Yes."
It just gets harder to say.

There is an interesting footnote that needs to be added to the discussion of the words "no" and "yes". In a well-run organization with a solid foundation and with supervisors who are creative, the higher a position is on the corporate ladder, the more often the person holding that position will have to say "no," and the more difficult it will be to say "yes." As a supervisor rises in an organization, he begins to supervise people who are supervisors themselves. When a supervisor, who is above the lowest step in supervision, is asked to address a problem, it means that it was too difficult for one, or more, of the supervisors who report to him to resolve. Since the first level supervisors were creative, not getting to "yes" means that a resourceful person could not resolve the problem. Suddenly the upper level supervisor is trying to resolve what the first level supervisor was not able to settle. With each step up the ladder, it becomes more difficult to be able to say "yes." The inability to say "yes" is because every supervisor below him has tried unsuccessfully to resolve the same issue.

If a supervisor has a need to feel appreciated by saying "yes," he should stay near the lowest level of supervision.

Out the Door
Where people sit in group meetings

There is a lot to learn by analyzing where people sit when they are allowed to select their own seats in a meeting. Actually, a person's choice of seats offers one of the clearest examples of body language. If the same group meets with some degree of regularity, people will consistently take the same chair. Associations can be noted because people will elect to be with the people for whom they have a natural affinity. Additionally, people who sit as far to the back as possible do so with the intention that they will not be noticed. In contrast, those who sit in the front center want to be recognized. Anyone standing next to the door wants to be on the outside of it.

The author's mentor took seat selection one step further, maintaining that there was a direct correlation between where people choose to sit when in large groups and where they stand politically (another spectrum). This seating observation is based on the viewpoint of the person holding the meetings. Those who sit to the right of the speaker tend to be the more conservative. Those who sit in the center were politically moderate, while those who sit to the left were more likely to be politically liberal (again this in the context of the group assembled). He took this example one step further, maintaining that the greater the distance a person sat from the leader in either direction, right or left, the greater the depth of their political views. Observation showed that the pattern was true in multiple organizations.

Humor as a measure of success
Recognize the quick mind: it can help.
Someone who has a quick witty response is almost always bright. A quick retort is very different from telling a joke. People have time to think of an appropriate joke, while a retort is best when served very fresh. Supervisors need to recognize those members of the staff who are known for their quick responses; their clever minds can be used to solve serious problems.

When is a joke not a joke?
The supervisor determines what is appropriate.
Supervisors set the standard for their organization. A smile or humor used to lighten the situation can be a great tool. Conversely, the trepidation of a situation is made even worse when a supervisor uses sarcasm.

The meeting pulled on the supervisor
Who's in control?
When an employee unexpectedly appears at a supervisor's door, the employee is creating a meeting. Since the employee has a reason for being at the door, he has set the agenda. The employee has also chosen the site and probably established the

time of day. The supervisor begins a meeting like this on the employee's terms. These "Oh, by the way," meetings can be the most difficult because control begins in the hands of the employee. The practice of an employee setting a meeting is especially a problem for supervisors who maintain an open door policy. It is suggested the supervisor consistently use *"walk and talk."*

Same Space – Different Messages
It is never easy.

An example of how the same space can be perceived differently is the office conference room. If the supervisor calls the meeting in the conference room, it is his space. If two employees meet there without the supervisor, it is a neutral space. The difference between the two examples is who is in the space.

Staff Development

"I don't care what they say. Some people are like pets; they can't be trained!"

Staff Development: the art of getting what you need and deserve.

Staff Development
If in doubt, train, then retrain and finally train again.

The value of staff development is often overlooked by both the organization and its employees. Staff development comes in two distinct forms, internal and external. Internal development (usually informal) is the passing of information by those who work for the organization. External development are those activities brought in such as workshops and seminars. The fact that the organization is spending money to train an employee should acknowledge his value and improve morale; however, this is rarely the case. Too often the employee sees training as an imposition or just a day out of the office. At the same time, organizations frequently envision staff development as an expense not an investment. When there is no clear focus or objective to the training both these perceptions are probably correct.

Essential components of staff development are providing employees with the training necessary to perform their responsibilities and the opportunity to refresh and refocus. When planned and managed, staff development is much more; it becomes the process by which an organization can change direction.

Organizations approach external staff development from two poles. When an organization determines in advance which seminar topics will be approved, or provided, based on its ongoing mission or its goals, staff development changes from being an expense to an investment. The other process is almost random and an expense. This is when each staff member is allowed to pick seminars, workshops and conventions that he wishes to attend.

The impact of the internal form of staff development, where advisors and mentors share information among the staff, is almost always overlooked. This internal component exists in every organization but is rarely managed.

The key to a successful staff development program is planning; including knowing advance what is going to be covered in workshops.

Staff Development
The Inertia of Organizations

Done correctly, staff development is the way to manage change. Ignored, or unsupervised, staff development is one of the ways organizations become stagnant or even regress.

Staff development creates inertia within organizations. In both scientific and organizational terms, inertia has two forms. There is inertia to have the organization remain as it is, and there is inertia that moves an organization forward. When staff development is designed to meet the goals of the organization, it can generate the energy needed to propel the organization forward. When an organization tries to save money by reducing staff development programs, it produces inertia at rest. When inertia is allowed to be at rest for too long, the organization becomes less competitive and often begins to implode.

Training versus Staff Development

The process of learning about the organization and an individual's position within that organization is essential for him to be successful. How, and from whom, one learns his job is a key factor in determining his level of success.

Training is the process of learning the skills directly related to the person's position. Formal training usually occurs when a person first starts a new position or when there are new duties added to his workload. Whether it is how to use the copier, or how to use new software, it is generally acknowledged that all employees need instruction in order to be effective. Training for a specific job is instruction that is specific and that is expected to be followed. Because it is specific most initial job training is individual in nature and done by the organization or by a vender that has sold the same system to multiple organizations. Only part of what an employee learns is formal and managed by his supervisor.

Organizations are good at training employees in the skills

needed for their positions but not always good at providing orientation – in "the way things work." The informal information, about the real priorities and internal dynamics, is usually acquired by inference and from the person's peers. Knowing how an organization "works" is not a skill, but it is required knowledge for a person to be successful. Because it is usually not part of the formal training program, counseling in the way things "work" is almost always completed informally by others within the organization. Supervisors need to be aware that if, as the person in charge, they do not set up a formal orientation process instructing people in "how things work," the employee will find (or be found by) the informal system.

After the initial employment training, staff development should take training to a higher level than simply improving specific skills. In fact, thought provoking staff development programs may not be directly related to the person's position. In thought provoking programs the employee is usually expected to determine how to modify the information so that it may be utilized in his position. Staff development in the higher form would include software training that shows a person how to modify the information to meet the organization's needs. When not training for a specific position, staff development can be conducted through programs that instruct groups comprised of people from several different organizations.

Informal Training

In every organization, training and staff development are completed both formally and informally. Formal training and development is planned, designed and required. Informal training and development is acquired through other people in the organization or through programs undertaken by the person of his own volition. Formal staff development is almost always positive, whereas informal training among employees is often not in the best interests of the organization.

Supervisors need to understand that people are always learning. If they are not learning through planned programs, they are learning through informal means and through their peers. Since all staffs are constantly undergoing some form of inertia, supervisors should plan to use the process of staff development to further the goals of the organization.

Gains through staff development - almost

Planned to assist in meeting the mission, staff development should benefit both the organization and its employees. The reality is that those benefits rarely materialize. Many employees see staff development as an obligation, or a necessary burden or even a day out of the office. At the same time, supervisors tend to have unreasonable expectations as to the impact of the development on the employees who attend.

Workshops tend to fall into two major categories they are either inspirational (meant to excite the employees) or informational (they disseminate information). People who attend inspirational workshops tend to leave feeling better or more enthusiastic but unless there is systemic change in the person within days the effect has worn off. Staff development that is informational is often prepackaged and not designed for the organization. If the information does not fit the needs of the organization, then it is like an expensive sweater that is too small, it looks good but the owner can never wear it.

The question becomes where is the disconnect between what should happen and what does happen? The answer is in the planning, the expectations and the follow-up.

In-house workshops

If an organization is bringing in a workshop, to train its own staff, it should establish in advance how the program will help meet its mission. The presenters need to understand what is expected and then modify the message to meet the needs of the organization. Outside presenters can not be faulted for using a packaged presentation, if the organization did not arrange in advance for specific points to be addressed. With the program designed for the organization, the staff needs to understand before they attend what the workshop is about, why they were chosen to attend and what changes are anticipated as a result. When these components are in place the development should have realist expectations.

The reason workshops do not cause long-term systemic change is that there was not a follow-up plan. Just sending someone to get information does not change his behavior. What he needs is a system that requires him to use the information from the workshop.

Off-site workshops and conventions

The problem with offsite workshops (these are workshops set up around a theme or topic with representatives from multiple organizations in attendance) is the message has to be general in nature. Since the topics are general these workshops require the person in attendance to determine how to use the information. Even with the restriction of a general message the level of application can be increased by insuring the person knows why they are attending, and what is expected when they return. Often these general workshops are shopping trips, where the person who attends is looking at the messenger to see if the presenter should be used by the company for an in-house workshop. Even with these off-site workshops there has to be a follow-up program or the message will be lost.

The formal process

In most organizations staff development, that is beyond skills training, consists of seminars, workshops, institutes and college courses, in which cases the leaders (presenters) are usually not employed by the organization (*five bridge rule*).

The Internal Staff Development Structure – the in-house

Internal staff development refers to training performed by people employed by the organization; however, not all internal development is planned, supervised or even accurate. It is the nature of organizations that staff members meet and while they are together they discuss their duties and the organization. It is through these informal conversations that information is transferred. Because many of these internal dialogs are unsupervised, there is a good chance that they may be unhealthy for the organization.

When an employee learns from others within the organization, the people from whom they learn fall under the roles of advisors, guides, mentors, or sages. Whether or not they are formally recognized, within every organization there are people who fit into each of these categories.

Advisors are everywhere. *Advisors* are those people who are all too willing to provide an opinion. There are *advisors* in a person's spiritual life, social relationships, education or work environment. *Advisors* are there, ostensibly, to help a person through a problem or situation. The reality is that *advisors* are so numerous that they are much easier to find than to lose.

Organizations consistently have a problem with *advisors* who are self-appointed. The seemingly open admission to the role of *advisor* leads to a lack of control over the quality of the advice being given. Some advice is excellent, but even more advice has limited value. This lack of control places the receiver in the position of judging the quality of the information. Judging the quality of advice is a difficult task for experienced employees, but, it is especially difficult for new employees. Supervisors need to be sure that, to the extent possible, employees turn to the right people for support. One way to help the staff find the right *advisor* is for the supervisor to indicate which staff member should be asked, if there are follow-up questions.

Guides are people who are recognized for their specialized knowledge or expertise; their advice is specific to a situation or problem. *Advisors* try to provide information in many areas; while *guides* are there to help with one specific area. There is a parallel between an outdoor guide and a professional *guide* within an organization. An outdoor guide may be an expert on mountain ranges, but may have no knowledge about deserts. A professional *guide* may be knowledgeable about software applications, but have no idea of how to help with sales. Within an organization, *guides* are usually the people in staff, not line, positions.

A **sage** is a person who has gained wisdom based on his experience and knowledge. The sage's wisdom allows him the unique ability to be almost always correct. Although a *sage* imparts his assistance to individuals, the nature of his advice supports the organization not the individual. When asked, the *sage* is willing to provide guidance to anyone in the organization, one time. If the *sage* is listened to with respect and his suggestions are heeded he will continue to support the person who asked. A *sage* is wise, so if his advice is ignored or, worse yet, scoffed at, he will stop sharing with that person. *Advisors* often go to other employees in an effort to convey their thoughts, the *sage* waits for people to come to him.

From the **mentor** to his *intern* is the way knowledge is imparted between generations. A *mentor* uses a situation as an opportunity to teach problem solving techniques to his *intern*. While a *guide* takes a person through the process required to solve one problem; a *mentor* tries to teach a system for resolving multiple problems. The *mentor's* role is to prepare the *intern* to handle future situations as they arise. *Guides* may work with several people, but a *mentor* works with only one person. The *mentor* is like an individual tutor who sees his role as that of a provider of skills that will help his *intern* long after the two have parted. The relationship between a *guide* and the person with whom he is working is based on the *guide's* knowledge. The relationship between the *mentor* and his *intern* is personal and is based on mutual trust and respect. *Mentors* are, universally, older than their *interns* and the *mentor* is at the same or a slightly higher professional level as the person he *interns*. Because of the potential for jealousy between *interns*, most *mentors* can only work with one *intern* at a time. The relationship between a *mentor* and his *intern* is not one-sided. The *intern's* interest provides validation (a form of praise) for the *mentor's* effort.

When an employee starts supervisors should look for someone to mentor the person. The supervisor should be careful not to be personally jealous of the relationship between the mentor and the intern. Most highly successful supervisors may have forgotten all their advisors and guides but they can name the people who served as their mentors.

Problems arise if there is a lack of control over this internal structure. When there are not clearly delineated "people to go to," learning takes place from whomever steps forward. When anyone is allowed to provide information, too often it is given by someone who is either not truly knowledgeable or who is disgruntled with the organization. If these malcontents are allowed to lead, internal staff development becomes detrimental to the organization.

Advisors are there to help anyone who will listen, with any problem that the advisor believes they have.

Guides are there to help multiple people with a single specific issue.

The sage serves the organization by helping anyone who seeks his advice.

For each intern there is only one mentor at a time. For each mentor there is only one intern at a time. Yet there is no deeper or more intense professional relationship than that of mentor/intern because of the level of trust and respect that they share for each other.

External Staff Development – the outhouse

Organizations should envision external staff development as corresponding to the internal structure. The models for external structure consist of: seminars, workshops, and institutes, parallel the internal structure of advisors, guides, and a combination of mentors/sages.

Seminars are comparable to the role of advisor. Seminars are plentiful, are generally good, and some are even practical, similar to the ideas passed on by advisors. Seminars are designed to expose a group of people to one idea. Because seminars have time constraints, and often have a large attendance, they tend to be either inspirational or informative. If the basis of the seminar is personal inspiration, the goal is to make those who attend want to enhance their performance. If the seminar is informational, it is designed to assure that those in attendance understand a regulation, benefit or something that is factual. The quality of a seminar should be measured by the number of the people who attend who are performing differently two weeks later.

Workshops are the guides of staff development. Designed to take several people through a process by practicing skills, workshops will usually provide a clear outcome. Like the guides in the internal structure, the presenter at a workshop was chosen because of his expertise. The impact of workshops lasts longer than that of seminars because the information is more specific and is practiced.

Institutes serve as the mentor/sages of staff development. Institutes are led by people who have turned their expertise and experiences into wisdom. The leader of an institute brings together a small group to go through a process for which there is not necessarily a clear answer. The desired outcome of an institute is not an answer, but an understanding of a process that will lead to the answers to multiple problems. Institutes have a second, often unintended, purpose; they provide an opportunity for supervisors to develop a peer group.

Virtually all **planned** staff development that is provided through reputable services is worth the time and effort. Staff development that is both planned and focused on an organization's established goals will help to create the *Critical Mass* necessary to institute change. There is little to be lost and much to be gained whenever planned and structured staff development is undertaken.

Because each person constantly grows, staff development is continuous. The question is, is his professional growth planned or does it have life of its own?

The best advice is: if in doubt, train, train and train some more.

The Caboose Effect:
There's always a last car

If you uncouple the last car from a long train, there will still be a last car, but training will keep that car moving with the rest of the train.

Supervisors all have days when they say, "If only (*insert name*) were gone, things would be so much better." In reality, (*insert same name*) probably is the biggest problem, but that person is seldom the only problem. Although there may be universal agreement among the staff that (*insert same name again*) performance is the weakest in the organization, there is rarely universal support for serious action against him.

There comes a time when each person ascribes most of what goes wrong to the performance of another person or to a small group – the blame game. When this happens, it is human nature to start focusing on the person's weaknesses, usually at the expense of ignoring or forgetting the person's strengths.

If (*insert the same name again*) actually left, within a week a different employee would be perceived as the weakest link. If the new weakest person left within a week, there would be yet another person who would be the weakest link. It is the same principle as uncoupling the last car from a long train. As the train pulls away from the uncoupled car, there is still a last car. Accepting that principle means that organizations need to stop trying to solve their problems by uncoupling employees, and start improving the performance of those people through training.

Once, as an assistant principal, I caught all four of the roughest boys in the school smoking together. These boys were far enough along on the progressive discipline chart that they each received a multiple-day suspension. Anticipating how pleasant the next three days were going to be, the teachers were relieved. At the end of the first day, I remember telling the principal that by the third day the teachers would be complaining about "x" and "y", two other boys who were far from good, but whose behavior was overshadowed by the first group. As predicted, by the third day the teachers were complaining about the two boys as discipline problems – the *caboose effect*.

There are positives outcomes when uncoupling (dismissal) the last car. If an organization discharges the weakest link, two outcomes become evident to those who remain. First, the organization will take a stand on performance; second, when the organization takes a stand, it can cost an employee his livelihood.

The downside to a dismissal is the unsettling effect on the remaining employees. There is an underlying understanding of the *caboose effect* that causes employees to rally around the weakest link. The other employees do not endorse the performance of the weakest; rather they realize that, if the weak person is gone, the supervisor will have more time to evaluate their own performances more thoroughly. In effect, the pecking order will be disrupted.

The *caboose effect* does not work at the front end of the train. Trains and organizations are pulled by energy. In the case of trains, that energy comes from engines. In the case of organizations, the energy comes from people, especially those who lead. When a car is lost from the back of the train, there is more energy for the remaining cars. However, when an organization changes leadership (the front cars), it reduces the strength in the engine required to pull the body. This energy loss can cause the organization to begin to implode.

The solution to the *caboose effect* is to make reasonable efforts to improve the performance of the weakest person before uncoupling occurs. The person needs to be told how to improve his performance and how to demonstrate the improvement according to a specified timeline. If he still holds back the train, uncoupling will be appropriate and understood.

Maintaining the last car is more efficient than trying to build a new one.

The Fastest Way to Learn
Is to study those whom you don't respect

Observation of others is a major part of learning to be a supervisor. Not surprisingly, knowledge is gained much faster when examining those we don't respect than by watching those we do.

People learn more quickly from those whom they do not respect, because it is much easier to identify something that is being done incorrectly than it is to ascertain why things go properly. When an aspiring, or new, supervisor sees what he believes to be a mistake made by another supervisor, he invariably notes it and says to himself, "I would never do that." Having seen the problem, the aspiring supervisor will probably not make that particular error in judgment. It is, however, much more difficult to identify the elements that make good supervisors successful.

Determining the faulty traits of those who are perceived to be unworthy of their supervisory position is relatively uncomplicated. If help is needed in determining a supervisor's weaknesses, there are always other people, usually hanging around the water cooler, who are willing to point out the person's shortcomings. Trying to avoid those same traits in one's own behavior is the challenge.

Supervisors are usually criticized by their staff members for the same traits that the supervisors are complaining about in the staff. The problems fall into the following general areas: lack of fairness in dealing with others, poor communications, expecting special privileges for himself, failure to be visible, disorganization, failure to take responsibility, inability to make a decision and mood swings.

It is more difficult to identify and to emulate the positive qualities of supervisors. The reason is simple – what people find offensive is narrow and easy to define, what people like is usually much broader and general and seems to be based on the individual's personality. For example, when people comment, "He never greets me." They are being specific, when they say

"He is always in a good mood," they are making a generalization. People can train themselves to recognize faults in others (the cure for the former); it is far more difficult to become an optimist (the recipe for the latter). A supervisor can avoid developing negative traits by observing other supervisors whom he does not respect; however, he can only maximize his potential by emulating those who are really successful.

There is a formula for improving as a supervisor. Each person knows instinctively how to watch a supervisor whom he does not respect and to break down that person's flaws into clearly defined traits. That provides one fourth of the growth formula. The second quarter of the formula is to train oneself not to display those same behaviors. The third quarter of the formula is to break down into measurable elements the positive qualities of those who are successful. The formula is completed by training oneself to emulate the positive behaviors demonstrated by successful supervisors.

Note: Half of the formula for improvement calls for training. The need for training applies to both supervisors and the staff. If in doubt, train, train and retrain.

What are the benefits of the workshop?
When asked to approve workshops supervisors should ask two questions. Do the individual programs fit into a planned goal or does each a stand alone? Remember that one of the keys to creating the critical mass is the selection of which staff attends which workshops.

Are the right people attending workshops?

Good at Many Things
or Master of One

The issue of (or with) consultants

Over the past twenty years there has been an ever-expanding expectation that supervisors should be able to handle more and more complex problems. Today's supervisors are facing issues with employees, systems, technology, financing, legal issues and government regulations. The diversity of the issues requires such a combination of knowledge and skills that it is virtually impossible for a supervisor to be an expert at any of them. Most organizations have addressed this problem by seeking supervisors who were generalists – good at many things, and utilizing consultants – experts at one thing.

What began as a small breed of experts has suddenly become a large herd of consultants (author included). This plethora of consultants is an essential resource; however, each one should be used for one purpose and only one purpose.

The truly bizarre situation exists when a consultant purports to be an expert in more than one area. Would you hire Einstein to be your spiritual leader, or a retired military officer to be your financial expert? If you answered "yes" put the book down, you are not ready for the rest. Assuming you said "no," why consider the same agency or person to be your consultant in more than one area.

Although the consultant's job is to help an organization become more successful, his motive is to stay in business and make a profit, not that there is anything wrong with that. A consulting firm, that works in only one area builds its reputation by helping its clients resolve problems in that one area. If the consulting firm is successful, other businesses employ them, and the firm attains its goal of making a profit.

You never know who you affect;
Or even how much.

As a supervisor, when a person does not agree with a decision, he is usually able to make sure that the message about his feelings gets back to the supervisor. There are, however, members of the staff who sit quietly observing and learning. Unfortunately, an observer will probably never tell his supervisor the positive effect he had on him. The supervisor's sense of justice and his fairness allowed him to serve as a role model. It is often years before this silent intern ever expresses his thanks. Even if he does not provide direct feedback the supervisor's example is often passed on through the observer's actions toward others.

A Consigliore
The rarest of all birds

In addition to advisors, guides, sages and mentors, there is one other role played by someone who is totally trusted by the head of an organization. This person serves as the consigliore or counselor to the head. The Consigliore's role is to provide guidance on major issues. Since the head needs to appear to be in control and the relationship is based on trust and confidentiality the consigliore is almost never identifiable. Because he is trustworthy the consigliore would not admit his role even if someone guessed. The consigliore is so hidden he may not even be present at meetings or even enter the building. The most common consigliore is a member of the board of trustees or the former head of the organization (but because of the nature of the relationship no one can do a study).

The Feeling of constantly giving
Everyone wants a piece of a good thing

Supervisors, especially those who make good decisions, find that they are constantly being asked more questions. The staff's logic is simple: "ask the person who will give the most appropriate (fair) answer." The problem that this logic creates is that the better the supervisor the less time he has to accomplish the task of his own position. It soon becomes obvious that good supervisors never have the luxury of 'down time.' Even worse, everyone who comes to their doors (and there are times when it looks like there is a line) wants an answer or "a little piece of them" Ultimately, the function of constantly providing for others has a wearing effect on the supervisor.

There is an answer: *Every supervisor needs to find someone, within the organization, with whom he can just visit.*

Note to the reader: The converse to this phenomenon is also true: supervisors who do not answer questions find themselves left alone. While having the ineffectual supervisor left alone may not serve the organization, it does serve him.

Reduced to a Pronoun
When people cannot stand each other

There is an interesting way to determine when the relationship between two people had degenerated to an unworkable level. When two people truly cannot accept each other, neither will refer to the other by name. In each case the antagonist has become a pronoun. The person is suddenly only "he or him," as if the use of a pronoun has robbed him of the dignity of a name. The problem for the supervisor is that because of the tension, even with only a pronoun to go on, the identity of the adversary is easy to determine.

Standards of Living Rarely Change
A person should take a position only if he wants to do the work.

There are at most, three times in a person's career when the change of positions truly changes his standard of living. To change one's standard of living something dramatic has to be altered. The ability to buy a second home, a move to another geographic area, are options that change one's standard of living. When considering a promotion, a person should ask himself, "What will change in my *personal* life if I accept this position?" People who already own two cars may be able to have two nicer cars (style) but they still have two cars. In most promotions the standard of living does not change, just the style of living.

If the standard of a person's personal life is not going to change, then promotions should be taken because the work is going to be more interesting or more challenging. In this case, the standard of living that has changed is in the person's professional life. Note: personal first, professional second.

A supervisor should take a position for the task involved; work takes too much of a person's life not to be enjoyed.

Getting Started

"Tomorrow we really should get something done."

It all begins sometime

Eight to Live by:

Never Hire Someone you can't fire

One's hire One's; Two's hire Three's

If in doubt, always hire the brightest

Pluralist lead us, Idealists steer us

People only question things they think they understand

Looking Brilliant let others try their ideas

Chip Theory
It isn't a game; it just looks like one

Everything Matters

Getting Started
Do in once; do it right

After reading a book filled with ideas on how to improve, the natural question is, "How do I get started?" The answer lies within each person; however, there are some suggestions for people who are undertaking the process alone.

1. Do nothing different tomorrow. It is better to make thoughtful changes than to rush in and make a mistake. If you try to make changes before you are ready, they may not work and it is likely that you will give up.
2. Tomorrow, find a person who you trust and who is available to discuss the ideas in the book. The ideal person should be someone who can act as your coach, as you act as his. The person need not be someone from work; it could be someone from the health club, neighborhood or car pool. The main criterion is that you trust his opinion.
3. Agree to read sections of the book at the same time. Select axioms (2 or 3) for which each of you will be responsible for leading a conversation. The conversation should begin with an axiom and its application to an experience in your own professional life. If you do not feel comfortable using examples from work, use the news. Every day there are stories of issues involving supervisors. [In this step an axiom is being applied to a situation.]
4. As the process continues, reverse the system and take an issue from work or the news and try to apply axioms to the event – this will make you analyze. [In this step a situation is being applied to an axiom.]
5. Still later, try to develop your own axioms based on multiple situations.
6. Early on, list behaviors in yourself that you want to change. Limit the number to a maximum of ten. This list needs to be work-related and realistic.

7. Develop a plan for changing these behaviors. You will need the plan for Step Nine.

8. Rank the behaviors from the easiest to change to the most difficult (one being the easiest).

9. Rank the same behavior a second time based on the estimated time required to change the behavior (one being the shortest amount of time).

10. For each behavior multiply the rank in eight by the rank in nine, i.e. if being more visible was ranked number three for ease and number two for time necessary it would have a product of six.

Compare the answers in number ten. The lower the product, the sooner that behavior should be addressed. In effect, the process has performed triage. Before the first plan is actually instituted, it is a good idea to talk over your findings with your coach. See if he agrees with your expectations. The most common mistake is having the wrong perception of how long a change will take to implement.

Successes are like bricks; they actually build on each other. As the accomplishments grow, a supervisor feels a greater sense of professional confidence. This confidence allows him to become more comfortable and more open to those who work with him.

Everything Matters
Ongoing

As the topics for **Everything Matters** were being compiled, more and more people became aware that a book was being written that was intended to help people to be better supervisors. Each person who learned of the book wanted to be sure that an incident in which he was involved was somehow included. The list of topics grew faster than the writing about the concepts they covered. As the months went by, it became obvious that, at the rate the writing was going, as compared to the adding of new ideas, the book **Everything Matters** would never be done. On one long walk, it suddenly occurred to me that *not* being complete was the whole point. The book was supposed to be finished but *Everything Matters* should never be concluded. After all *Everything Matters* is a process – not a project – and like all good processes, it should continue as long as it works. After the walk, the focus became to present the concept and enough examples and ideas to allow others to continue the process of analysis and synthesis to create axioms. The objective had shifted from presenting a collection of axioms to showing how they are used and encouraging people to develop their own.

Keep the Process of Creating Axioms Going

The cycle of successful supervision is anticipate, plan, respond, review, and refresh. Anticipate problems in advance. Plan a strategy to avoid the problem. If a problem does surface, respond appropriately. As soon as possible after the problem is resolved, review why the problem was not avoided and review the appropriateness of the response. On a regular basis take a break and refresh. Although all incidents are part of the natural process of *Everything Matters*, one of the easiest ways to implement this cycle is through examination of meetings. Those in attendance should ask themselves: "What came up?" "How could it have been avoided?" and "What was the response required?"

With conscientious practice all people can learn to analyze problems by breaking them down into their component parts. The process of putting the parts from multiple situations back together (synthesis) into a usable shape is more complicated. The process of synthesizing data becomes more fun when one tries to create his own axioms. The assurance that the new axiom works comes when at a later time, it can be applied to a new situation. These "proven" axioms serve as quick reminders of the lessons learned in previous situations.

Ben Franklin's Dollar

American statesman, Ben Franklin, was born to a humble family. Through his own efforts, he became a very wealthy man. There is a story that, at one point, a young boy asked him for the loan of a dollar, promising to pay Franklin back when he became successful. After lending the boy the money, Franklin went to France, where he lived for several years. After Franklin returned to the United States, he was walking down a street in Philadelphia when the young boy, who was now a man, rushed over and tried to repay Franklin the money he had borrowed years before. Franklin, a wise and wealthy man, said to the boy, "Give the money to someone else starting out. Just tell the person who you give it to that you are passing on Ben Franklin's dollar."

Years ago, while serving as my mentor, Jerry Murphy passed the dollar to me. Through this book I am paying back my Ben Franklin's dollar – now my debt is resolved. It is now up to the reader and my interns to continue the process.

Also by Deep Roots Publications
A collection of Victorian Era true crimes that prove
truth is stranger than fiction.
Visit us on the web at www.deeprootspublications.com.

In six months she went from being the mistress of a man who was nominated for governor to a person charged in a double homicide. But with her conviction came a curse on the lawyers involved in the trial.

A collection of six stories in which it is known who committed the crimes but they all got away because of the Rules of Victorian society.

An Irish maid whose murder rattled a small village, but not as much as the suicides, five years later of four successful male friends.

A collection of three stories. From a bank robbery to mass murder, these stories are based on the trials that made the lawyers famous enough to be in the Billings's trial.

Although the evidence was weak, the wealthiest man in a very rich county, Jesse Billings, was charged with the murder of his wife. This was one of the first trials covered by the national press.

<u>Everything Matters: Ideas for Supervisors</u> is the first in a trilogy. Keep posted on the status of other books by visiting our website at www.everythingmattersthebook.com.